Theory and Applications in Mathematical Physics

Conference in Honor of
B. Tirozzi's 70th Birthday

Theory and Applications in Mathematical Physics

Conference in Honor of
B. Tirozzi's 70th Birthday

Editors

Elena Agliari
"Sapienza" Università di Roma, Italy

Adriano Barra
"Sapienza" Università di Roma, Italy

Nakia Carlevaro
ENEA, C.R. Frascati, Italy

Giovanni Montani
ENEA, C.R. Frascati;
"Sapienza" Università di Roma, Italy

World Scientific

NEW JERSEY · LONDON · SINGAPORE · BEIJING · SHANGHAI · HONG KONG · TAIPEI · CHENNAI · TOKYO

Published by

World Scientific Publishing Co. Pte. Ltd.
5 Toh Tuck Link, Singapore 596224
USA office: 27 Warren Street, Suite 401-402, Hackensack, NJ 07601
UK office: 57 Shelton Street, Covent Garden, London WC2H 9HE

Library of Congress Cataloging-in-Publication Data
Names: Tirozzi, Brunello. | Agliari, Elena, 1980–
Title: Theory and applications in mathematical physics : in honor of B. Tirozzi's 70th birthday /
    edited by Elena Agliari (Sapienza Universitá di Roma, Italy) [and three others].
Description: New Jersey : World Scientific, 2015. | Includes bibliographical references.
Identifiers: LCCN 2015028012 | ISBN 9789814713276 (hardcover : alk. paper)
Subjects: LCSH: Mathematical physics. | Statistical mechanics.
Classification: LCC QC20 .T4435 2015 | DDC 530.15--dc23
LC record available at http://lccn.loc.gov/2015028012

British Library Cataloguing-in-Publication Data
A catalogue record for this book is available from the British Library.

Printed in Singapore

B. Tirozzi

(*Photo by Dino Ignani*)

Preface

The conference for my 70th birthday has been very amusing for me. I had the possibility to meet very nice colleagues with whom I shared a nice time of my life, both for the research we have done and for the company. Also for them it was an occasion to meet each other. I have to say that the enjoyment had always been an important part in my life. When I speak about enjoyment I mean to have fun making research about some interesting topics in physics or mathematics or both. I have always found that physics is full of very attracting topics and that treating them with some good mathematical formalism was even more exciting. Thus my scientific career has always been based on the search of the arguments that at that moment looked to me interesting and attracting. So I did not work all the time on the same topic, many young researchers remain stuck on the same argument all the time for being sure that they will produce a lot of papers. I think that this is not a good choice.

I started my activity at the Department of Physics at the University "La Sapienza" in 1968. It was a very hot time. There was an intense activity of research of the theoretical group in many directions and there was an intense political activity at the same time. I was among the young people who participated to all the rallies, meetings, occupations but I took part also in the research activity which was also very extraordinary. I took the degree on physics with Marcello Cini as supervisor and we worked on the electromagnetic mass of the elementary particles, application of group theory to particle physics. The coworkers where Franco Bucella, who continued in this direction and Mimmo De Maria who decided later to dedicate himself to the history of physics. But at the same time there were lot of seminars on strong interaction among elementary particles, Regge poles, analytic structure of scattering amplitudes, perturbation theory and so on. Just to mention the main persons I cite Nicola Cabibbo, Luciano Majani, Francesco Calogero, Tony de Gasperis, Gianni Jona Lasinio, Giuliano Preparata, Carlo Di Castro. But the methods applied by these brilliant theoretical physicists were not so satisfactory for me because theoretical physics often hides the hypothesis and the assumptions at the base of their theories, or make some approximations that are not clearly written. So I got interested in the rigorous derivations made by Gianni Jona and Francesco Calogero. But my first approach to rigorous mathematical physics started with the group founded at that time by Giovanni Gallavotti.

A lot of young people participated to his seminars and started to develop interesting applications of probability theory to statistical mechanics, quantum systems, dynamic of infinite systems of particles. At that time I produced very good papers in collaboration with Errico Presutti, Mario Pulvirenti and others. So we discovered that the right formulation of statistical physics can be done using probability theory. On those years, around the '70, very important and nice publications appeared on Russian journals, and being translated in English. I am speaking about the papers by Yasha Sinai, Roland Dobrushin, Bob Minlos. But at that time there was also the cold war, the iron curtain, no internet and so we received these important papers one or two years later. These scientists were masters of applications of probability theory to phase transitions since they were all coming from the school of Kolmogorov. I wanted also to see the life in Soviet Union and to understand how they constructed the communist society. With these two strong curiosities, I make the big move to go to Soviet Union for one year using the exchange program among CNR and the Academy of Science. I got a very warm welcome by the group of Sinai, Dobrushin and Minlos. Of course I had to learn Russian since their talks were mainly in Russian. The discussion at their seminars were even more intense than in our seminars. I remember that it was not possible to speak more than 3 minutes without being interrupted by a lot of questions. Since all the big scientists were not allowed to move due to the iron curtain, Moscow was full of very good seminars held by other well known people. I quote the Gelfand's seminar, Katok's seminar and many others. So I learned Russian and probability theory at the same time. The life in Soviet Union was much less exciting than the scientific activity I must say. When I came back I brought some techniques that were useful for all the group. After that year I often went to work with this exciting group in Moscow. In one of these contacts I got acquainted with some rigorous result obtained by Sinai and Khanin about the spin-glass. This result was used in a very good paper I have written with Enzo Olivieri and Marzio Cassandro. I have also generalized a nice result about periodic orbits by Sinai and Vul and this was my first paper where I used the computer. In the '82 I also explored other ways of doing research, I always remained in contact with the Russian group but I was curious to see how mathematical physics was done in other part of the world. So I decided to ask the hospitality of Joel Lebowitz at Rutgers University who had a big group of statistical physics. I had the experience to teach a course on differential equations there.

In the second part of that year I went to visit IHES (the Institute of Hautes Etudes Scientifiques) in Paris. Of course the atmosphere in Paris was very exciting and the scientific activity was at a very high level. The director was in fact David Ruelle. At this institute I started a very interesting collaboration with Kristoff Gawedzky and Antti Kupiainen with whom we published very good papers on renormalization group. On that time I was commuting between IHES and the Mathematical Department in Rome. In that time string theory came up very strongly and in fact there were seminars everywhere and also at IHES and in Rome. I took part in an interesting seminar about string theory held by Claudio Procesi, De Concini, Enrico Arbarello and the physicists Massimo Testa and Giancarlo Rossi. But somehow, even if I have published a paper with De Concini and Fucito I was not satisfied with this theory because there was no experimental counterpart. That is why I again changed topic and came back to statistical physics but at this time I was fascinated by the Hopfield model which incorporated the property of memory, associative memory, in the statistical physics landscape. The model also describes in a simple way the properties of the neurons. I created a new course at the Math. Department named Reti Neurali (Neural Networks) and many students attended my lectures. So I got a big number of enthusiastic young disciples. Some of these remained at the University or in the research. Our relationships were always very constructive and we helped each other very much. I would like to mention the following: Enrico Ferraro, Enrico Rossoni, Silvia Puca, Sara Morucci, Daniela Bianchi, Giulia Rotundo, Gabriele Stabile, Marco Piersanti. With them we had experience of different applications of neural networks. With Enrico Ferraro and Enrico Rossoni we modeled the behavior of the oxytocin neurons, with Giulia Rotundo and Gabriele Stabile we made different models of markets and economy, with Silvia Puca, Sara Morucci and Stefano Pittalis we applied neural networks to the wave motions and tide motions of the sea. The oxytocin problem was coming from a European collaboration organized by Jianfeng Feng about neurobiology. Daniela Bianchi was a very precious coworker for studying biology and for the study of the network of the neurons of the hyppocampus made with the NEURON program, very difficult to manage, especially if trying to use it on parallel computers. Marco Piersanti was also another very good student of mine who entered in this research.

The problem of the motion of the Italian seas was coming from a contract of a government agency. I met Janfeng Feng in Beijing University where he invited me for one month. I was also very interested to visit Beijing. We started a very nice collaboration about neural networks which went on for many years with a rich production of papers. In those years I went also many times to Bochum, Bielefeld, Bonn where I worked with Sergio Albeverio and his group getting very interesting results. But I had also the fortune to be contacted on those times by Masha Shcherbina and Leonid Pastur of Kharkov Ukraine. They are leaders in the application of the probability theory to the Hopfield model. We arrived to solve rigorously the Hopfield model without using the non-rigorous replica trick used by Amit, Gutfreund and Sompolinsky for solving the model. We also continued with many other interesting rigorous results. I found a very good fellow in the Math. Department and after in the Physics Department who shared with me the desire of solving the disordered models with rigorous probability theory. I am speaking about Francesco Guerra who has shown very important results with rigorous methods about the Parisi's solution of the replica symmetry breaking of spin glass theory. Our researches were going on in parallel and it was encouraging for me to talk with some person about this approach because nobody had the courage, at that time, to go through such problems. For two reasons, the first was that it was very difficult to find rigorous proofs, the second was that the physicists solving their models with the replica theory were doing opposition against our approaches. Francesco has helped me a lot in this struggle and also helped me on the practical level. At the end both of us won this battle and got the prizes we were looking for.

In Italy I became a member of CINFAI, an Interuniversity Consortium for studying the Physics of Atmospheres and Hydrospheres. I was interested to study this topic. Sergio Albeverio organized a meeting for me with Sergey Dobrokhotov of Moscow. He is applying the asymptotic methods developed by P. Maslov to the problem of propagation of typhoons and tsunami. We started a very fruitful collaboration on these arguments with very original results. This collaboration still goes on, fortunately and we also succeed to use asymptotic methods for the problem of Plasma in the Tokamak. By the law of the Italian government I am obliged to retire in 2014. But again my collaboration with Russian groups was lucky for me because the Plasma Laboratory of Frascati showed interest in these new methods and I got the hospitality at their center. So at the end of all my going around in the world and in the science I am still in the active research. I think that this is a consequence of my tendency to follow only the things that looked exciting to me.

I retire from the University to start another life in the research. I am glad to see here all my best companions of my long life in the science. I am also very happy to have here many of my best disciples and students. Further, above all, I have to thank one million times Adriano Barra and Elena Agliari who organized the heavy work of this lucky conference. There is also Biancamaria with whom we shared a lucky life and other relatives. So I have just to say goodbye, the story is not finished but continues with some changes.

Thanks to all.

B. Tirozzi

Contents

Chapter 1

Motifs stability in hierarchical modular networks

E. Agliari

Dipartimento di Matematica, "Sapienza" University of Rome (Rome, Italy)

A. Barra

Dipartimento di Fisica, "Sapienza" University of Rome (Rome, Italy)

F. Tavani

Dipartimento di Scienze di Base e Applicate per l'Ingegneria (SBAI), "Sapienza" University of Rome (Rome, Italy)

B. Tirozzi

Dipartimento di Fisica, "Sapienza" University of Rome (Rome, Italy) and Enea Research Centre Frascati

Recent advances in our understanding of information processing in biological systems have highlighted the importance of modularity in the underlying networks (ranging from metabolic to neural networks), as well as the crucial existence of *motifs*, namely small circuits (not necessary loopy) whose empirical presence in these networks is statistically high. In these notes, mixing statistical mechanical with graph theoretical perspectives and restricting on hierarchical modular networks, we analyze the stability of key motifs that naturally emerge and we prove that all the loopy structures have systematically a broader steadiness with respect to loop-free motifs.

Introduction

Network theory, coupled with Statistical Mechanics, is becoming a crucial tool for investigating Biological Complexity and, following this approach, crucial questions, ranging from intra-cellular investigations (as for instance in metabolic or protein networks[1,12,16,17]), to extra-cellular ones (as for instance in neural networks[4,9,20,22]) have already been satisfactorily addressed.

According to empirical evidence, biological networks typically exhibit scale-free and/or hierarchical topologies[18,23,24] with a high number (with respect to a random reference) of "motifs", namely recurrent and strongly-connected sub-graphs or patterns.[19] Further, the interaction strength (e.g., based on lock-and-key mechanisms[4]) between the elements making up the network usually varies over several orders of magnitude, in accordance with a log-normal or power-law distribution for link's magnitudes. This has stimulated a renewed interest for the Dyson model:[14] indeed the latter, originally developed as a model to overcome mean-field limitations in the statistical mechanical description of ferromagnetism, is exactly a hierarchical network, where spins are pasted on its nodes and their couplings follow a power-law distribution.[8] In this abstract model spins may play as neurons (thus dealing with a hierarchical neural networks[3]) and one is interested in controlling a clique of firing neurons in a sea of quiescent ones, or may be thought of as lymphocytes (i.e. clones of B or T cells), thus dealing with hierarchical immune networks,[9] and one may question the activation of a cluster of clones while all the others are silenced and so on.

In these notes we aim to analyze the meta-stability of key motifs highly occurring in the Dyson model, as their existence has been found only recently.[2,3] In particular, we consider the *dimer*, i.e., the prototype of a loop-less reticular animal, and the *square*, i.e., the prototype of a loopy reticular animal, and we check whether magnetic configurations where spins associated to these motifs are misaligned with respect to the bulk -but aligned among themselves- are stable. Not surprisingly, while the former is found to be always unstable (i.e. there is no value of the tuneable parameters defining the model that allows its stability), the latter has a range of stability. It is worth noting, however, that -as these motifs are by definition not-extensive (i.e. their sizes do not scale with the system size)- nor they contribute to the model's free energy in the thermodynamic limit, neither they are expected to be stable whenever a finite-amount of fast noise is applied on the system.

As a last remark, we note that the Dyson model has a power law distribution for the link's magnitude[8] as well as a modular architecture of the graph hosting the spins:[3] remarkably, the reason for the stability of its loopy motifs lies exactly in these intrinsic features of the model, that -in turn- play a major role even in real biological networks,[15] exactly those where the presence of motifs is expected.[18,19]

1. Definition of the model

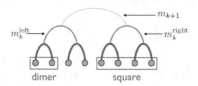

Fig. 1.   Schematic representation of the hierarchical network defined by the Dyson model of size 4. Each node hosts an Ising spin and interactions among nodes are stronger (here denoted by thicker links) for closer nodes as depicted in the Hamiltonian (eq.(1)) defining the model.

The aim of this section is to give a microscopical description of the Dyson Hierarchical Model (DHM), which is composed by 2^{k+1} Ising spins S_i, for $i = 1, \cdots, 2^{k+1}$, that are embedded in a hierarchical topology. The model is represented by the Hamiltonian introduced in the following definition:

Definition 1. The Hamiltonian of Dyson's Hierarchical Model (DHM) is defined by

$$H_{k+1}(\vec{S}|J,\sigma) = H_k(\vec{S_1}) + H_k(\vec{S_2}) - \frac{J}{2^{2\sigma(k+1)}} \sum_{i<j=1}^{2^{k+1}} S_i S_j, \qquad (1)$$

where $J > 0$ and $\sigma \in (1/2, 1]$ are numbers tuning the interaction strength. Clearly $\vec{S_1} \equiv \{S_i\}_{1 \leq i \leq 2^k}$, $\vec{S_2} \equiv \{S_j\}_{2^k+1 \leq j \leq 2^{k+1}}$ and $H_0[S] = 0$.

Thus, in this model, σ triggers the decay of the interaction with the distance among spins, while J uniformly rules the overall intensity of the couplings. Note further that the coupling distribution $P(J)$ is scale free as it follows the power-law relation:[8] $P(J) \propto J^{-\frac{1}{2\sigma}}$. We can introduce the partition function $Z_{k+1}(\beta, J, \sigma)$ at finite volume $k + 1$ as

$$Z_{k+1}(\beta, J, \sigma) = \sum_{\sigma}^{2^{k+1}} \exp\left[-\beta H_{k+1}(\vec{S}|J,\sigma)\right], \qquad (2)$$

and the related free energy $f_{k+1}(\beta, J, \sigma)$, namely the intensive logarithm of the partition function, as

$$f_{k+1}(\beta, J, \sigma) = \frac{1}{2^{k+1}} \log \sum_{\vec{S}} \exp\left[-\beta H_{k+1}(\vec{S}) + h \sum_{i=1}^{2^{k+1}} S_i\right]. \quad (3)$$

We introduce also the global magnetization $m = \lim_{k\to\infty} m_{k+1}$ where

$$m_{k+1} = \frac{1}{2^{k+1}} \sum_{i=1}^{2^{k+1}} S_i, \quad (4)$$

that can be defined recursively, level by level (see Figure 1). Finally, we denote the thermodynamical average as

$$\langle m_{k+1}(\beta, J, \sigma)\rangle = \frac{\sum_{\vec{S}} m_{k+1} e^{-\beta H_{k+1}(\vec{S}|J,\sigma)}}{Z_{k+1}(\beta, J, \sigma)}, \quad (5)$$

and $\lim_{k\to\infty}\langle m_{k+1}(\beta, J, \sigma)\rangle = \langle m(\beta, J, \sigma)\rangle$.

We are interested in understanding the conditions to be applied on σ such that different configurations remain stable in noiseless conditions. We will start with some simple cases, and we will try to apply the results to a general structure composed by 2^n elements, with $n < k + 1$.

2. Loop-less case: Stability analysis of the dimer

The goal of this section is to study the existence of possible values of σ such that the dimer (i.e. a spin-configuration where $S_i = +1$ for $i = 1, 2$ and $S_j = -1$ for $j = 3, \cdots, k + 1$) remains stable, clearly in the noiseless limit. To reach our conclusions, we adapt to the case an interpolative strategy -firstly developed in[13]- that has been recently applied to hierarchical networks:[2]

Definition 2. Once considered a real scalar parameter $t \in [0, 1]$, we introduce the following interpolating Hamiltonian

$$H_{k+1,t}(\vec{S}) = -\frac{Jt}{2^{2\sigma(k+1)}} \sum_{i>j=1}^{2^{k+1}} S_i S_j$$

$$- \frac{Jm(1-t)}{2^{(2\sigma-1)(k+1)}} \sum_{i=1}^{2^{k+1}} S_i + +H_k(\vec{S_1}) + H_k(\vec{S_2}), \quad (6)$$

such that for $t = 1$ the original system is recovered, while at $t = 0$ the two-body interaction is replaced by an effective, tractable one-body term. The possible presence of an external magnetic field can be accounted simply by adding to the Hamiltonian a term $\propto h \sum_i^{2^{k+1}} \sigma_i$, with $h \in \mathbb{R}$.

This prescription allows defining an extended partition function as

$$Z_{k+1,t}(h, \beta, J, \sigma) = \sum_{\vec{S}} \exp\{-\beta[H_{k+1,t}(\vec{S}) + h \sum_{i=1}^{2^{k+1}} S_i]\}, \qquad (7)$$

where the subscript t stresses its interpolative nature, and, analogously,

$$\Phi_{k+1,t}(h, \beta, J, \sigma) = \frac{1}{2^{k+1}} \log Z_{k+1,t}(h, \beta, J, \sigma). \qquad (8)$$

It is easy to show that

$$\Phi_{k+1,0}(h, \beta, J, \sigma) = \Phi_{k,1}(h + mJ2^{(k+1)(1-2\sigma)}, \beta, J, \sigma), \qquad (9)$$

so that we can write

$$\Phi_{k+1,1}(h, \beta, J, \sigma) = \Phi_{k+1,0}(h, \beta, J, \sigma) + \int_0^1 \frac{d\Phi}{dt} dt \implies$$

$$\Phi_{k+1,1}(h, \beta, J, \sigma) = \Phi_{k,1}(h + mJ2^{(k+1)(1-2\sigma)}, \beta, J, \sigma) + \int_0^1 \frac{d\Phi}{dt} dt. \,(10)$$

Using the identity (10), with the appropriate computations, we obtain

$$\Phi_{k+1,1}(h) = \Phi_{k+1,0}(h) - \frac{\beta J}{2}(2^{(k+1)(1-2\sigma)}m^2 + 2^{-2(k+1)\sigma})$$

$$+ \frac{\beta J}{2}2^{(k+1)(1-2\sigma)} \left\langle (m_{k+1}(\vec{S}) - m)^2 \right\rangle_t$$

$$\geq \Phi_{k,1}(h + Jm2^{(k+1)(1-2\sigma)}) - \frac{\beta J}{2}(2^{(k+1)(1-2\sigma)}m^2$$

$$+ 2^{-2(k+1)\sigma}). \qquad (11)$$

We already know that we can study non-standard stabilities where the system undergoes the influence of two different contributions m_1 and m_2, with the same absolute value, but opposite in sign.[2] Now we want to analyze the case in which the subsystems have different cardinality: in this case, the first one is constituted only by two spins, and the other one is constituted by all the others; we can write the two sub-magnetizations m_1 and m_2, more rigorously, as:

$$m_1 = \frac{2}{2^{k+1}} \sum_{i=1}^{2} S_i, \qquad m_2 = \frac{2^{k+1} - 2}{2^{k+1}} \sum_{i=3}^{2^{k+1}} S_i, \qquad (12)$$

where m_1 has an opposite sign with respect to m_2. This means that the system preserves the same general magnetization m up to the last level, where one has 2^k blocks formed by 2 spins: at this point, one of the blocks, say the one formed by S_1 and S_2, is not affected by the rest of the system, while all the others 2^{k-1} blocks continue to interact each others. Rigorously, starting from the formula (11) and iterating the same interpolating estimates up to the first level (i.e. the level in which we are considering spins at distance $d_{ij} = 1$) we arrive at

$$\Phi_{k+1,1}(h) \geq \Phi_{1,0}(h + Jm\sum_{l=2}^{k+1} 2^{l(1-2\sigma)}) - \frac{\beta J}{2}(\sum_{l=2}^{k+1} 2^{l(1-2\sigma)}m^2 + \sum_{l=2}^{k+1} 2^{-2l\sigma}).$$
(13)

Now, asking for dimer stability, we assume that

$$\Phi_{1,0}(h + Jm\sum_{l=2}^{k+1} 2^{l(1-2\sigma)}) = \frac{1}{2^k}\Phi_{1,0}^1(h + Jm\sum_{l=2}^{k+1} 2^{l(1-2\sigma)})$$

$$+\frac{2^{k+1} - 2}{2^{k+1}}\Phi_{1,0}^2(h + Jm\sum_{l=2}^{k+1} 2^{l(1-2\sigma)}), \text{(14)}$$

where we wrote the two contributions to the interpolating free energy from different groups of spins.

With some straightforward calculations we obtain that

$$\frac{1}{2^k}\Phi_{1,0}^1(h + Jm\sum_{l=2}^{k+1} 2^{l(1-2\sigma)})$$

$$= \frac{1}{2^k}\Phi_{0,1}^1(h + m_1 J 2^{1-2\sigma} + mJ\sum_{l=2}^{k+1} 2^{l(2\sigma)})$$

$$= \frac{1}{2^k}\log\cosh\left[\beta J(m_1 2^{1-2\sigma} + h + m\sum_{l=2}^{k+1} 2^{l(1-2\sigma)})\right]$$

$$+\frac{1}{2^k}\log 2,$$
(15)

and this allows writing

$$f_{k+1} \geq \log 2 + \frac{1}{2^k} \log \cosh \left[\beta J(m_1 2^{1-2\sigma} + h + m \sum_{l=2}^{k+1} 2^{l(1-2\sigma)}) \right]$$

$$+ \frac{2^{k+1} - 2}{2^{k+1}} \log \cosh \left[\beta J(m_2 2^{1-2\sigma} + h + m \sum_{l=2}^{k+1} 2^{l(1-2\sigma)}) \right]$$

$$- \frac{\beta J}{2} (\sum_{l=2}^{k+1} 2^{l(1-2\sigma)} m^2 + \sum_{l=2}^{k+1} 2^{-2l\sigma})$$

$$- \frac{\beta J}{2} 2^{1-2\sigma} \left(\frac{2m_1^2 + (2^{k+1} - 2)m_2^2}{2^{k+1}} \right). \tag{16}$$

Since we are looking for conditions on σ for the stability of the dimer, to ensure that the term with m_1 is much more significant than the term with m in the logarithm, we need to fulfill

$$\sum_{l=2}^{k+1} 2^{l(1-2\sigma)} < 2^{1-2\sigma}. \tag{17}$$

In the thermodynamic limit of the system (i.e. when $k \to +\infty$), this means

$$\sum_{l=2}^{\infty} 2^{l(1-2\sigma)} < 2^{1-2\sigma} \iff \frac{2^{2(1-2\sigma)}}{1 - 2^{1-2\sigma}} < 2^{1-2\sigma} \iff 2^{1-2\sigma} < 1 - 2^{1-2\sigma} \iff$$

$$2^{2-2\sigma} < 1 \iff 2 - 2\sigma < 0 \iff 2\sigma > 2 \iff \sigma > 1. \tag{18}$$

The condition (18) shows that the dimer can be stable only if $\sigma > 1$, but this bound violates the condition $\sigma \in (1/2, 1]$, as stated in the definition of the model. This result can be expressed in the following

Theorem 1. *We can not find conditions on σ such that a misaligned dimer remains stable in the noiseless limit.*

Remark 1. It is possible to observe that the first term on the right of the (16) goes to zero as k tends to infinity. This means that in the thermodynamic limit the contribute of the dimer becomes negligible in the total free energy as it should.

3. Loopy case: Stability analysis of the square

Here we focus on the structure made by spins whose configuration is $S_i = +1$ when $i = 1, \cdots, 4$ and $S_j = -1$ when $j = 5, \cdots, 2^{k+1}$, and we will see that now there exist some possible values of σ such that this configuration remains stable.

We recall that the mean-field interpolating Hamiltonian can be written as in (6), and the partition function and the free energy are respectively like in (7) and (8). Proceeding as already done for the dimer, using (9) and (10), we can imagine that all the spins in the system maintain the same magnetization up to the second level, and then we have two different contributions: the first one given by a group composed by only four spins, and the second one given by all the other ones. Then, in this case, we can write

$$m_1 = \frac{2^2}{2^{k+1}} \sum_{i=1}^{4} S_i \qquad m_2 = \left(\frac{2^{k+1} - 4}{2^{k+1}}\right) \sum_{i=5}^{2^{k+1}} S_i. \qquad (19)$$

With the same computations explained for the dimer, considering this new configuration, we arrive at

$$f_{k+1}(h, \beta, J, \sigma)$$

$$\geq \log 2 + \frac{4}{2^{k+1}} \log \cosh(\beta J(m_1 \sum_{l=1}^{2} 2^{l(1-2\sigma)} + h + m \sum_{l=3}^{k+1} 2^{l(1-2\sigma)}))$$

$$+ \frac{2^{k+1} - 4}{2^{k+1}} \log \cosh(\beta J(m_2 \sum_{l=1}^{2} 2^{l(1-2\sigma)} + h + m \sum_{l=3}^{k+1} 2^{l(1-2\sigma)}))$$

$$- \frac{\beta J}{2}(\sum_{l=3}^{k+1} 2^{l(1-2\sigma)} m^2 + \sum_{l=3}^{k+1} 2^{-2l\sigma}) - \frac{\beta J}{2} \sum_{l=1}^{2} 2^{l(1-2\sigma)} \left(\frac{m_1^2 + m_2^2}{2}\right). (20)$$

Again, investigating the values of σ for the stability of the square, we can state the following theorem

Theorem 2. *A misaligned square remains stable in the thermodynamic limit when* $\sigma \in (\frac{3}{4}, 1]$.

Proof. As shown in the previous sections, we start by asking

$$\sum_{l=3}^{k+1} 2^{l(1-2\sigma)} < \sum_{l=1}^{2} 2^{l(1-2\sigma)},$$

to ensure that the contribution given by $m_{1,2}$ is more significant than the one given by m. In the thermodynamic limit the previous inequality turns out to read as

$$\sum_{l=3}^{+\infty} 2^{l(1-2\sigma)} < \sum_{l=1}^{2} 2^{l(1-2\sigma)}.$$

By computing this sum we have that

$$\frac{2^{3(1-2\sigma)} - 2^{(k+2)(1-2\sigma)}}{1 - 2^{1-2\sigma}} < 2^{1-2\sigma}(1 + 2^{1-2\sigma}),$$

that is $\sigma > 3/4$.

Since $\sigma \in (\frac{1}{2}; 1]$ by definition, differently from the dimer configuration, we found an interval of values of σ such that the square remains stable.

□

We can conclude that, moving from the dimer to the square, we find crucial differences: while the first configuration can never be stable in the thermodynamic limit, the second one -under appropriate conditions on σ- remains stable. It is worth noticing that the model free energy is obtained summing over all the contributions, thus even those from both the subsystems, however -in the thermodynamic limit- their contributions are vanishing as expected, thus (finite-size) motifs, while crucial for dynamical processes on networks, do not contribute to its thermodynamics.

4. Generalizations to other loopy motifs

In the previous sections we have proved that loopy motifs can be generically stable in hierarchical networks with scale-free distributed couplings, while less structured motifs without loops in general can not. Scope of the present section is to extend the previous results toward more general loopy motifs, showing that the procedure used to find the critical value of σ, can be extended to find the stability of other configurations. In particular, we are going to analyze the general case where

$$m_1 = \frac{2^n}{2^{k+1}} \sum_{i=1}^{2^n} S_i, \qquad m_2 = \frac{2^{k+1} - 2^n}{2^{k+1}} \sum_{i=2^n+1}^{2^{k+1}} S_i, \qquad n \in [1, k+1]. \quad (21)$$

Proceeding exactly as for the square configuration, we imagine that at the n-th level, with $n \in [2, k]$, the two groups, one constituted by 2^n spins, and the other one by $2^{k+1} - 2^n$ have different magnetization. Following step-by-step the general scheme and calculations outlined in the previous sections we can finally state

Theorem 3. *The general configuration of 2^n spins, where they are arranged according to eq. 21, remains stable in the thermodynamic limit when $\sigma \in (\frac{n+1}{2n}, 1]$ with $n \in [1, k]$.*

Remark 2. The condition $\frac{n+1}{2n} \leq 1$ is verified for all $n \geq 1$.

Remark 3. Via the route paved in these notes we can infer the range of validity of σ in a novel way. Indeed imposing that $n = k$, and then taking the limit $k \to \infty$, this results in the condition that $\sigma \in (\frac{1}{2}, 1]$.

5. Conclusions and outlooks

At a *macroscopic scale*, from a graph theory perspective, several biological networks are scale-free and show modularity,[15,17,20,21] while, at the *microscopical scale*, these networks are characterized by the presence of highly expressed reticular motifs.[19,23]

In these notes, driven by a statistical mechanical guide, we analyzed the stability of motifs within the Dyson model,[14] as the latter can be read as a hierarchical network where modules naturally emerge and whose distribution of coupling strength is scale-free.[2,8]

Of course, the exhaustion of motif's exploration -already confined within this model- is out of the scope of the present notes as here we restricted the analysis to the prototype of a loop-free structure, that is the *dimer*, and to the prototype of a loopy structure, that is the *square*, while tracing out the general guidelines to be followed to generalize further investigations on Dyson network's motifs at will.

In particular, we successfully proved that, while the loop-free motif always lacks stability and will always fluctuate, for the latter -the loopy motif-there is a region of stability (in the space of the tuneable parameters), in agreement with intuition and with previous results from the Literature.
Next efforts will be devoted to a systematic exploration of network's motifs within the modular-hierarchical as well as within more general scale-free networks.

Acknowledgments

A.B. acknowledges Laboratories for Information Food and Energy through the project "INNOVA-MATCH". F.T. is partially supported by "Avvio alla Ricerca 2014", Sapienza University of Rome that is acknowledged too.

References

1. E. Agliari, A. Barra, R. Burioni, A. Di Biasio, G. Uguzzoni, *Nature Sci. Rep.* **3**, 3458, (2013).
2. E. Agliari, A. Barra, A. Galluzzi, F. Guerra, D. Tantari, F. Tavani, *J. Phys. A* **48**, 015001 (2015).
3. Agliari, A. Barra, A. Galluzzi, F. Guerra, D. Tantari, F. Tavani, *Phys. Rev. Lett.* **114**, 028103 (2015).
4. E. Agliari, A. Barra, A. Galluzzi, F. Moauro, F. Guerra, *Phys. Rev. Lett.* **109**, 268101 (2012) .
5. E. Agliari, A. Barra, A. Galluzzi, F. Guerra, F. Tavani, D. Tantari, *Neur. Nets.* **66**, 22 (2015).
6. E. Agliari, A. Barra, A. Galluzzi, D. Tantari, F. Tavani, *NCTA2014: Neural computation theory and application*, 210, (2014).
7. E. Agliari, L. Asti, A. Barra, R. Burioni, G. Uguzzoni, *J. Phys. A* **45**, 365001, (2012).
8. E. Agliari, A. Barra, A. Galluzzi, D. Tantari, F. Tavani, *Phys. Rev. E* **91**, 062807 (2015).
9. A. Annibale, A. Barra, P. Sollich, D. Tantari, *Phys. Rev. Lett.* **113**, 238106 (2014).
10. A. Barra, A. Bernacchia, E. Santucci, P. Contucci, *Neur. Nets.* **34**, 1 (2012).
11. A. Barra, F. Guerra, G. Genovese, D.Tantari, *JSTAT* P07009, (2012).
12. L.D. Bogarad, M.W. Deem, *Proc. Natl. Acad. Sc. USA* **96**, 2591 (1999).
13. M. Castellana, A. Barra, F. Guerra, *J. Stat. Phys.* **155**, 211 (2013).
14. F. Dyson, *Commun. Math. Phys.* **12**, 91, (1969).
15. L.H. Hartwell, J.J. Hopfield, S. Leibler, A.W. Murray, *Nature* **402**, C47 (1999).
16. L. Dello Schiavo, M. Altavilla, A. Barra, E.Agliari, E. Katz, *Nature Sci. Rep.* **5**, 9415 (2015).
17. H. Jeong, B. Tombor, R. Albert, A.L. Barabasi, *Nature Lett.* **407**, 651 (2000).
18. S. Maslov, K. Sneppen, *Science* **296**, 910 (2002).
19. R. Milo, et al., *Science* **298**, 824 (2002).
20. P. Moretti, M.A. Munoz, *Nature Comm.* **4**, 2521, (2013).
21. E. Ravasz, A.L. Somera, D.A. Mongru, Z.N. Oltvai, A.L. Barabasi, *Science* **297**, 1551 (2002).
22. B. Tirozzi, D. Bianchi, Introduction To Computational Neurobiology And Clustering, *World Scientific Publ*, (2007).
23. A. Wagner, D.A. Fell, *Proc. R. Soc. Lond. B* **268**, 1803 (2001).
24. R. Weiss, et al., *Nat. Comp.* **2**, 47 (2003).

Chapter 2

Asymptotic techniques in the solution of the Maxwell equations in plasmas

A. Cardinali

ENEA - Fusion Technical Units, R.C. Frascati (Frascati (Rome), Italy)

The propagation of the electromagnetic modes from low (Radio Waves) to high frequencies (TeraHertz) in a magnetized (or unmagnetized) plasma is described by means of the Maxwell's equation system coupled to the plasma dynamics that can be modeled by the Vlasov equation. Although a fluid description in the plasma dynamics simplifies considerably the theory, a global solution would still be excluded due to the complexity of the matter. In some relevant conditions, after linearizing the equation system, an asymptotic treatment of the problem can be given in terms WKB[4–6] expansion of the fields. At the lowest order, a non linear first order partial differential equation for the Phase Integral, formally equivalent to the Hamilton-Jacobi equation in classical mechanics, can be obtained and solved in terms of the ray trajectories, while at the next order a transport equation for the slowly varying wave energy density can be obtained and solved, thus allowing to reconstruct the electric field inside the plasma. Examples of solution for the propagation of the Lower Hybrid Waves (LHW), relevant in the heating and current drive of laboratory plasmas confined in toroidal devices oriented to the research on thermonuclear fusion (tokamak), and the propagation of Whistler Radio Signal in the upper ionosphere will be shown and discussed.

1. Introduction

Historically the mathematical technique kwown as the WKB method (named after physicists Wentzel, Kramers, and Brillouin, who all developed it in 1926) can be traced back to the nineteenth century (see Ref. 1 for a discussion on the historical development of the WKB method). Modern developments of the theory started in 1915 (Gans,[2] Jeffreys[3]). In 1926 Brillouin,[4] Kramers[5] and Wentzel,[6] introduced the method in Quantum Mechanics to treat the Schrdinger wave equation. Extensions of the the-

ory have been developed in many aspects and are closely related to the
higher order geometrical optics. In general the method was normally applied for approximating the solution of a differential equation whose highest
derivative is multiplied by a small parameter. The electromagnetic wave
propagation in plasmas is described by a partial differential equation system that comes from the Maxwell's equations together with the constitutive
relations. The plasma, including the effect of the imposed static magnetic
field, is in general an inhomogeneous and anisotropic medium.

Extensive discussions on this subject may be found in books like Budden,[7] Stix,[8] Ginzburg[9] and Brambilla.[10] The propagation of waves in the
above-mentioned medium may be described by either first order or second
order coupled wave equations. Using Maxwell's equations and the constitutive relations (relation between the current and the electromagnetic field),
which are based on the solution of the Vlasov equation, a non-linear integro-
differential equation system can be derived whose solution meets serious difficulties to be given. In many cases the model can be simplified by reducing
the system to a more manageable one, in particular (i) linearization of the
Vlasov equation (by considering the electromagnetic field as a perturbation
of the background) ii) Maxwellian assumption of the background (thermal
equilibrium), (iii) cold plasma limit, and (iv) plasma in steady state condition (time harmonic electromagnetic field); all these assumptions allow to
obtain a system of second order coupled linear partial differential equations
instead of the non-linear integro-differential one which, although impossible
to solve directly, can be treated with much more simplicity. One of the most
important methods of solution of the coupled wave equations governing the
wave propagation in inhomogeneous plasma is the WKB solution, which
has been firstly used by Clemmow and Heading[11] and Budden and Clemmow[12] in the theory of radio waves propagation in the ionosphere. The
propagation of electromagnetic waves in inhomogeneous plasmas and the
corresponding WKB solution has also been discussed in a general approach
by Ginzburg,[9] Budden,[7] Born and Wolf,[13] Wait,[14] Weinberg,[15] Bernstein[16]
and Friedland and Bernstein[17] and many others. Application of the WKB
method to laboratory plasma oriented to sustain the research on the thermonuclear fusion has been given in the past by Baranov and Fedorov[18]
Brambilla and Cardinali[19] in the range of frequencies named "Lower hybrid" ($\approx$ GHz), while Bornatici et al.[20] assessed the theory in the case
of "electron cyclotron" ($\approx$ tenth of GHz) wave propagation. In the case
of radio wave propagation in the upper ionosphere the WKB method was
applied to study the conversion of a launched electronic whistler to a lower

hybrid in order to explain some features of the measured signal (Campo-reale et al.[21]). In this article we will propose the WKB solution of the electromagnetic wave equations in a cold plasma (fluid approach) by deriving the reduced set of equations at the zeroth (phase) and first order (amplitude) in the WKB expansion parameter and in a general geometry. The Landau (and/or collisional) damping of the propagating wave will be included in the study by considering it in the equation for the amplitude at the first order of the expansion. Successively, as examples of the application of the method, we discuss the propagation of the Lower Hybrid Wave in a tokamak device (toroidal geometry) and the propagation and attenuation of a radio signal (electronic whistler) in the upper ionosphere, and eventually its conversion in a quasi-longitudinal Lower Hybrid. A particular attention will be devoted to describe the singular points (caustics) that during the propagation can be met by the wave and to discuss of the failure (and the eventual remedy) of the WKB technique in reconstructing the field on these points. The article is organized as follow: In Sec. 2, the general integro-differential Maxwell-Vlasov equation system is discussed and the simplifying hypothesis are introduced and discussed, leading to the relevant system of equations describing the propagating wave in dispersive medium. In Sec. 3 the WKB technique is illustrated and applied to the wave equation; a discussion of the limit of validity of the technique is also given. In Sec. 4 application of the theory to the propagation of the "Lower Hybrid Wave" in a laboratory plasma confined in the "tokamak" device with a toroidal geometry is presented. In the same section the propagation in the upper ionosphere of a "whistler" radio signal, radiated by a dipole antenna, is also calculated in the spherical earth geometry and the eventual mode conversion of the whistler in a quasi-longitudinal LHW is studied when considering density striation of the plasma in the earth magnetic field. Finally in Sec. 5 conclusions are drawn.

2. Pertinent equation

The Maxwell-Vlasov system of equations is the pertinent model in describing the propagation and absorption of an electromagnetic wave propagating in a dispersive absorbing medium such as that of laboratory plasma or that existing in natural way in the upper terrestrial atmosphere (100-1000 Km) called "ionosphere". Indeed, the interest of HF waves in plasma comes from ionospheric and astrophysical observations. It is worth to recall the experiment of G. Marconi (Nobel prize in 1909), who was sending a radio

signal (λ=1800 m) from Cornwall to Terranova (Canada, 1901), in defying the skepticism of the scientific community. In fact, following the theory of the rectilinear propagation of the electromagnetic waves, the signal was unable to follow the terrestrial curvature. In spite of the negative advise of the scientific community the experiment was successful. Appleton and collaborators discovered in 1925, above the terrestrial atmosphere a region which had the characteristics of an ionized gas (ionosphere) that was able to explain the non-rectilinear propagation of the radio signal, and therefore to explain the Marconi experiment (Appleton, Nobel prize in 1947).
The pertinent equation in describing the wave propagation is

$$\nabla \wedge \nabla \wedge \vec{E}(\vec{r},t) + \frac{1}{c^2}\frac{\partial^2 \vec{E}(\vec{r},t)}{\partial t^2} = -\frac{4\pi}{c^2}\frac{\partial \vec{J}(\vec{r},t)}{\partial t}. \tag{1}$$

The point now is to correlate J to E (constitutive relation) via a conductivity tensor (2nd rank tensor) that takes into account the medium characteristics. This can be done in a general form by

$$\vec{J}(\vec{r},t) = \sum_\alpha q_\alpha \int f_\alpha(\vec{r},\vec{v},t)\vec{v}d\vec{v}, \tag{2}$$

where the evolution of f is ruled by the Vlasov kinetic equation (VKE) for the self-consistent field (Ref. 10 for a clear derivation of the VKE)

$$\partial_t f_\alpha + \vec{v}\cdot\nabla f_\alpha + \frac{q_\alpha}{m_\alpha}\left[\vec{E}(\vec{r},t) + \frac{\vec{v}\wedge\vec{B}(\vec{r},t)}{c}\right]\cdot\nabla_v f_\alpha = 0. \tag{3}$$

Maxwell's equations together with the VKE constitute a closed system: *It is the fundamental set of equations for the description of HF waves in plasmas.* The Maxwell-Vlasov (MVKE) system is a 2nd order, non-linear integro-differential system of equations. Instead of dealing with the evident difficulties of facing of Eqs. (1-3), we can adopt a description of the plasma dynamics based on the fluid theory. In this case the constitutive relation Eq. (2) can be replaced by

$$\vec{J}(\vec{r},t) = \sum_\alpha q_\alpha n_\alpha \vec{v}_\alpha(\vec{E}(\vec{r},t)), \tag{4}$$

where the first two moment equations of the plasma hydro-dynamical description, the mass and momentum conservation, are used to correlate the fluid velocity to the electromagnetic field:

$$\partial_t \rho + \nabla\cdot(\rho\vec{v}) = 0$$

$$\rho\frac{d\vec{v}}{dt} = \frac{\vec{J}\wedge\vec{B}}{c} + q_\alpha n_\alpha\vec{E}, \tag{5}$$

where $\rho = \sum_\alpha m_\alpha n_\alpha$, $\vec{v} = \rho^{-1} \sum_\alpha m_\alpha n_\alpha \vec{v}_\alpha$ and $\vec{B}$ is the (eventual) external magnetostatic field. It is worth to note that the fluid model is characterized by the (i) impossibility to describe the collisionless absorption (Landau Damping), while the collisional absorption can be included by retaining in the momentum conservation equation the friction term ($\approx i\nu$) where ν is the electron-electron (e-e) and electron-ions (e-i) collision frequencies, (ii) absence of the pressure term (cold plasma limit); (iii) impossibility to treat "thermal waves" (e.g. Ion/Electron Bernstein waves[22]). The fluid approach can be useful in deriving a relevant set of equations which can be "mathematically" treated, provided that (i) linearization of the equations (the wave energy density much weaker then the thermal energy of the plasma); (ii) harmonic field perturbation ($\vec{E}(\vec{r}, t) = \vec{E}(\vec{r}) \exp(-i\omega t)$). This approach can be very useful to describe the propagation of Whistler, Lower Hybrid, Electron and Ion Cyclotron Waves. In general, without harmonic field hypothesis, the equation system (1)-(4)-(5) can be written compactly as

$$
\nabla \wedge \nabla \wedge \vec{E}(\vec{r}, t) = -\frac{1}{c^2} \left\{ \frac{\partial^2 \vec{E}(\vec{r}, t)}{\partial t^2} + \sum_{\alpha = e, i} \omega_{p\alpha}^2(\vec{r}) \left[\vec{E}(\vec{r}, t) \right. \right.
$$
$$
\left. \left. + \Omega_{c\alpha} A e^{\Omega_{c\alpha} A t} \left(\int_t e^{-\Omega_{c\alpha} A t} \vec{E}(\vec{r}, t') dt' + C' \right) \right] \right\}, \tag{6}
$$

where $\omega_{p\alpha}(\vec{r})$ and $\Omega_{c\alpha}(\vec{r})$ are respectively the plasma and cyclotron frequencies. The boundary conditions for the field are intended put on the boundary surface enclosing the plasma volume, and imposed by the current circulating in the antenna conductors. This equation is a vector integro-differential equation of hyperbolic type, which comes from the linearized Maxwell-Fluid model. The electromagnetic waves can be excited by external antennas, which can be simulated by current in external conductors. Note that $e^{\Omega_{c\alpha} A t}$ is the "matrix exponential" which depends on the structure of the external magnetostatic field.

As stressed above, a great simplification can be obtained when considering the harmonic field perturbation $e^{(-i\omega t)}$. The wave equation Eq. (6) reduces to the "Vector Helmholtz Equation":

$$
\nabla \wedge \nabla \wedge \vec{E}(\vec{r}, \omega) = -\frac{\omega^2}{c^2} \left(\underline{\underline{I}} - \frac{i}{\omega} \sum_\alpha \omega_{p\alpha}^2(\vec{r}) \underline{\underline{C}}_\alpha^{-1}(\vec{r}, \omega) \right) \cdot \vec{E}(\vec{r}, \omega), \tag{7}
$$

where $C_{=\alpha}^{-1}$ is the matrix

$$C_{=\alpha}^{-1} = \frac{1}{-i\omega(\omega^2 - \Omega_{c\alpha}^2)} \cdot$$

$$\begin{pmatrix} (i\omega)^2 + \Omega_{c\alpha}^2(\hat{b}_1)^2 & -(i\omega\Omega_{c\alpha}\hat{b}_3) + \Omega_{c\alpha}^2(\hat{b}_1\hat{b}_2) & (i\omega\Omega_{c\alpha}\hat{b}_2) + \Omega_{c\alpha}^2(\hat{b}_1\hat{b}_3) \\ (i\omega\Omega_{c\alpha}\hat{b}_3) + \Omega_{c\alpha}^2(\hat{b}_1\hat{b}_2) & (i\omega)^2 + \Omega_{c\alpha}^2(\hat{b}_2)^2 & -(i\omega\Omega_{c\alpha}\hat{b}_1) + \Omega_{c\alpha}^2(\hat{b}_2\hat{b}_3) \\ -(i\omega\Omega_{c\alpha}\hat{b}_2) + \Omega_{c\alpha}^2(\hat{b}_1\hat{b}_3) & (i\omega\Omega_{c\alpha}\hat{b}_1) + \Omega_{c\alpha}^2(\hat{b}_2\hat{b}_3) & (i\omega)^2 + \Omega_{c\alpha}^2(\hat{b}_3)^2 \end{pmatrix}$$

$$(8)$$

and $\hat{b}$ is the versor along the external magnetostatic field. By introducing the dielectric tensor

$$\epsilon_{=}^{DT}(\vec{r}, \omega) = \left(I_{=} - \frac{i}{\omega} \sum_{\alpha=e,i} \omega_{p\alpha}^2(\vec{r}) C_{=\alpha}^{-1}(\vec{r}, \omega) \right), \qquad (9)$$

we can write compactly

$$\nabla \wedge \nabla \wedge \vec{E}(\vec{r}, \omega) + \frac{\omega^2}{c^2} \epsilon_{=}^{DT}(\vec{r}, \omega) \cdot \vec{E}(\vec{r}, \omega). \qquad (10)$$

3. Solution of the Vector Helmholtz equation by asymptotic techniques

The solution of the equation by asymptotic expansion (WKB approximation), indeed simplifies the treatment of the equation, but can be applied only when considering waves whose wavelength is much shorter than the typical scale length variation of the macroscopic plasma parameters: Complications can arise owing to Geometry of the medium which contains the plasma (in this case the differential operators must include the metrics); Boundaries & Reflection of waves from the boundaries; Presence of singularities (cut-offs, mode-conversion, geometrical reflection layers); Dependence of the macroscopic plasma parameters from time (in general the plasma can be considered at the steady state, also the laboratory-plasma); Interest is, in relevant cases (for example RF applications to tokamak or propagation of whistler waves in ionosphere), for unbounded plasmas, toroidal geometry (for example plasma contained in a "tokamak device"), spherical geometry (ionosphere plasma) etc.

The WKB expansion of the field (WKB Ansatz) can be written as

$$\vec{E}(\vec{r}, t) = \vec{A}_0(\vec{r}) \exp\left[ih^{-1}\wp(\vec{r}) - i\omega t \right], \qquad (11)$$

where $\vec{A}_0(\vec{r})$ is the amplitude and $\wp(\vec{r})$ is the phase; $\vec{A}_0(\vec{r})$ and $\wp(\vec{r})$ are functions varying on different space scale: the phase varies on a much faster scale with respect to the amplitude. Using the Ansatz Eq. (11) in Eq. (10) we have at the lowest order $O(h^{-2})$

$$-\vec{A}_0(\vec{r})\left[\hat{\nabla}\wp(\vec{r})\right]^2 + \vec{A}_0(\vec{r}) \cdot \hat{\nabla}\wp(\vec{r})\hat{\nabla}\wp(\vec{r}) + \underline{\underline{\epsilon}}^{DT}(\vec{r}) \cdot \vec{A}_0(\vec{r}) = 0, \qquad (12)$$

where the operator $\hat{\nabla} = c/\omega\nabla$. This equation is satisfied only if

$$H \equiv \det\left[-\left(\hat{\nabla}\wp(\vec{r})\right)^2 \underline{\underline{I}} + \hat{\nabla}\wp(\vec{r})\hat{\nabla}\wp(\vec{r}) + \underline{\underline{\epsilon}}^{DT}(\vec{r})\right] = 0. \qquad (13)$$

This equation is formally similar to the Hamilton-Jacobi equation in classical mechanics, where H is the Hamiltonian and S is the Hamilton's principal function or action. It is a nonlinear first order partial differential equation and its solution gives information on the evolution in space of the phase integral. To find the slow variation of the amplitude $\vec{A}_0$ at the next order $O(h^{-1})$ is not a simple task. The starting wave equation Eqs. (10) is a system of partial differential equations of the second order. WKB applied to second-order vector or, equivalently, matrix differential equations has been studied in some simple situations by D.R. Smith,[23] and R. Spigler and M. Vianello.[24] Here for the sake of simplicity and considering that the Lower Hybrid wave is a quasi-longitudinal wave, we can study the amplitude evolution of the electrostatic wave equation obtained by taking the divergence of Eqs. (10). We obtain a scalar equation for the scalar potential

$$\nabla \cdot \left(\underline{\underline{\epsilon}}^{DT}(\vec{r},\omega) \cdot \nabla\phi(\vec{r},\omega)\right) = 0. \qquad (14)$$

The Amplitude Transport Equation is obtained by assuming the same ansatz Eq. (11) for the scalar potential

$$\phi(\vec{r},t) \approx \phi_0(\vec{r}) \exp\left[ih^{-1}\wp(\vec{r}) - i\omega t\right]. \qquad (15)$$

If we recast Eq. (14) as

$$\nabla\phi \cdot \left[\nabla \cdot \left(\underline{\underline{\epsilon}}^{DT}\right)^T\right] + \left[\left(\underline{\underline{\epsilon}}^{DT}\right)^T \cdot \nabla\right] \cdot \nabla\phi \approx \left[\left(\underline{\underline{\epsilon}}^{DT}\right)^T \cdot \nabla\right] \cdot \nabla\phi = 0, \quad (16)$$

where we have neglected the space derivatives of the dielectric tensor, we obtain the following equation for the scalar potential

$$\nabla\phi_0(\vec{r}) \cdot \left[\left(\underline{\underline{\epsilon}}^{DT}\right)^T \cdot \nabla\right]\wp(\vec{r}) + \phi_0(\vec{r})\left[\left(\underline{\underline{\epsilon}}^{DT}\right)^T \cdot \nabla\right] \cdot \nabla\wp(\vec{r})$$
$$+ \nabla\wp(\vec{r}) \cdot \left[\left(\underline{\underline{\epsilon}}^{DT}\right)^T \cdot \nabla\right]\phi_0(\vec{r}) = 0. \qquad (17)$$

This equation is a first order partial differential equation system and in order to be solved, it requires the preliminary solution of Eq. (13) for the phase function.

Equation (13), as prescribed by the Hamiltonian theory, can be solved by the method of characteristics. The ray equations are formally similar to the Hamilton equation, and can be written as

$$\dot{r} = \partial_{\vec{p}} H(\vec{r}, \vec{p}, \omega),$$
$$\dot{p} = -\partial_{\vec{r}} H(\vec{r}, \vec{p}, \omega), \tag{18}$$
$$d\wp = \vec{p} \cdot d\vec{r}.$$

Equations (18) are a coupled system of first order ordinary differential equations for the position $\vec{r}$ and the wave-vector $\vec{p}$ (conjugate variables), its numerical (or analytical) integration with the initial conditions given on the initial surface $\wp(\vec{r})|_\Gamma = \wp_0$ allows the reconstruction of the wave phase inside the plasma. In the next section we will give a solution of Eq. (18) in toroidal geometry (tokamak device) when the external magnetic field has only two components the toroidal and the poloidal (screw pinch). Note that the solution of Eq. (18) can be given only when the second derivatives of the phase function are determined. This means that, being the phase function evaluated numerically, numerical derivatives of the phase function or their gradient must be provided with extreme accuracy.

4. Numerical results for Lower Hybrid Wave in laboratory plasmas (tokamak devices) and the whistler wave in ionosphere

In this section we report some numerical results of the propagation of an electromagnetic pulse in the LHW range of frequencies in a toroidal magnetic confinement structure (tokamak). Results of the propagation of a whistler radio signal in the upper ionosphere in the earth magnetic field will also be presented. In order to reconstruct the phase function $\wp(\vec{r})$, we will solve Eqs. (18) in a tokamak geometry. The natural choice of the coordinate system is the toroidal one (r, θ, ϕ) where r is the radial variable, θ and ϕ are the poloidal and toroidal angles respectively. The metric coefficients are $h_r = 1$; $h_\theta = r$; $h_\phi = R_0 + r\cos\theta$, where R_0 is the major radius of the torus. In the case of whistler propagation in the ionosphere the natural geometry is the spherical (geographical) one (r, θ, ϕ) , where $r = R_c + r_h$ is the radius of the sphere ($r = 0$ is the Earth center, R_c is the Earth radius and r_h is the width of the ionosphere corona), θ is the polar angle (latitude) and ϕ the azimuthal angle (longitude). The metric coefficients are $h_r = 1$; $h_\theta = r$; $h_\phi = r \cos\theta$. The dispersion relation H can be easily calculated from Eq. (8) by knowing the elements of the cold plasma dielectric tensor (see T. H. Stix[8] and M. Brambilla[10]). In the Stix reference frame we have

$$H = An^4 + Bn^2 + C = 0. \tag{19}$$

The coefficients A, B, C are

$$\begin{aligned}
A &= S \sin^2\vartheta + P \cos^2\vartheta \\
B &= (S^2 - D^2)\sin^2\vartheta + PS(1 + \cos^2\vartheta) \\
C &= P(S^2 - D^2)
\end{aligned} \tag{20}$$

and S, D, P are the cold plasma dielectric tensor

$$S = 1 - \sum_{\alpha=e,i} \frac{\omega_{p\alpha}^2}{\omega^2 - \Omega_{c\alpha}^2}$$

$$D = \sum_{\alpha=e,i} \frac{\Omega_{c\alpha}}{\omega} \frac{\omega_{p\alpha}^2}{\omega^2 - \Omega_{c\alpha}^2} \tag{21}$$

$$P = 1 - \sum_{\alpha=e,i} \frac{\omega_{p\alpha}^2}{\omega^2}.$$

ϑ is the angle between the external magnetic field and the wave-vector, ω, $\omega_{p\alpha}$, $\Omega_{c\alpha}$ are the applied, the plasma and the cyclotron frequencies respectively, and in Eq. (19) the wave-number $\vec{n} = c/\omega\vec{p}$ has been introduced. Equation (19) is a bi-quadratic algebraic equation for $n = f(\vec{r}, \omega)$. At fixed ω, Eq. (19) defines two branches of space dispersion of the electromagnetic waves each with a different polarization: i) the "slow branch" (referring to the phase velocity) with E_z/E_x polarization and ii) the "fast branch" with E_y/E_x polarization. By using the dispersion relation Eqs. (19-21) as Hamiltonian we can solve the system Eqs. (18) by keeping several rays starting from an initial surface $\wp_0$ at the plasma edge which can be identified as an initial wave-front of the propagating wave. In 3D this wave-front will be a strap of the toroidal surface, that is illuminated by the antenna. In the case of the Lower Hybrid propagation in tokamak the antenna is a rectangular wave-guide that is overlooking the plasma edge. In the case of the radio signal "Whistler" propagation in the ionosphere the antenna is a dipole antenna and the initial phase surface S_0 (wave-front) is spherical. In general the problem related to the propagation of the LHWin a tokamak and the Whistler Wave in the ionosphere can be reduced to a bi-dimensional problem because all the plasma parameters like density, magnetic field etc.

depend only from 2 space coordinates, the third being ignorable. This leads to a trivial integration of one of the equations Eqs. (18), and the determination of a constant value for one component of the wave-vector. In the case of toroidal geometry the ignorable variable is the angle ϕ because the tokamak plasma density depends on r (plasma radius) and the external magnetic fields from r and θ (the poloidal angle) and this leads to $\dot{p}_\phi = -\partial_\phi H(r, \theta, \omega) = 0 \Rightarrow p_\phi = const$. The same occurs in the case of the spherical geometry where ϕ (the longitude) is the ignorable coordinate. In this case in fact the density depends only on r (the radial variable) while the earth magnetic field depends only on r and θ (the latitude). In Fig. 1 a plot of rays starting from a strap of toroidal surface $r = a$ (plasma minor radius), $-45° \leq \theta \leq +45°$ and for $\phi = 0$ is shown together with the equiphase surface $\wp(r, \theta) = const$. The rays reach the plasma center and suffer a radial reflection ($\dot{r} = \partial_{p_r} H(r, p, \omega) = p_r \partial_p H(r, p, \omega) = 0 \Rightarrow p_r = 0$) before coming back at the plasma surface. Note that a caustics forms at the reflection point where all the rays are tangent at that surface. In Fig. 2 is shown the value of the phase $\wp(r, \theta) = const$ in 3D plot. In Fig. (3,4) a plot of the caustic surface is shown in a poloidal section (3) and in 3D space (4); on the same plot (4) the magnetic axis of the torus is shown (black line). In Fig. 5 a 3D plot of the rays starting at the FTU antenna ($-45° \leq \theta \leq +45°$ and $-2° \leq \phi \leq +2°$) is given showing that the rays are propagating in space making a very big excursion in toroidal direction and remaining sufficiently collimated during the propagation. In Figs. 6 and 7 finally the surface $n_r(r, \theta) = (c/\omega)p_r(r, \theta)$ and $p_\theta(r, \theta)$ are plotted in space together with the equi-surface lines $n_r(r, \theta) = const$ and $p_\theta(r, \theta) = const$.

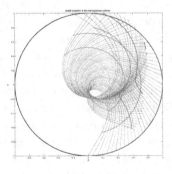

Fig. 1. Ray propagation in the poloidal plane

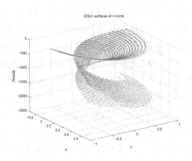

Fig. 2. Equi-phase surfaces

Fig. 3.   Caustics in the poloidal plane

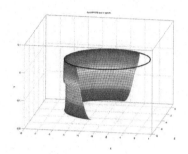

Fig. 4.   Caustics in 3D space

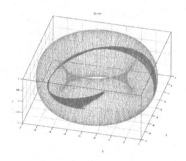

Fig. 5.   Ray propagation in 3D space

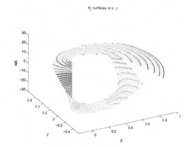

Fig. 6. n_r surfaces

In order to compute the slow evolution of the electric field amplitude for the LHW in a tokamak, we can specialize Eq. (17) in toroidal coordinate, in presence of a poloidal and toroidal component of the confinement magnetic field $B(r,\theta) = \sqrt{B_\theta^2(r,\theta) + B_\phi^2(r,\theta)}$. We got the following equation

$$p_r \frac{\partial \phi_0(r,\theta)}{\partial r} + \left[\frac{\alpha(r,\theta)p_\theta}{Sr^2} - \frac{\gamma(r,\theta)p_\phi}{2SrR} \right] \frac{\partial \phi_0(r,\theta)}{\partial \theta}$$

$$+ \frac{1}{2} \left[\frac{\partial p_r}{\partial r} + \frac{\alpha(r,\theta)}{Sr^2} \frac{\partial p_\theta}{\partial \theta} + \frac{\lambda p_\theta}{Sr} + \frac{\eta p_r}{Sr} + \frac{\mu p_\phi}{SR} \right] \phi_0(r,\theta) = 0,$$

$$(22)$$

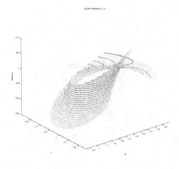

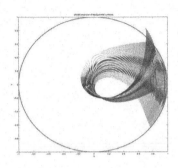

Fig. 7. m_θ surfaces   Fig. 8.   Field amplitude

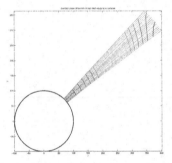

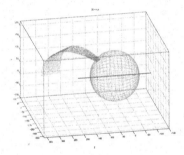

Fig. 9.   Whistler wave propagation in the   Fig. 10.   Whistler wave propagation in
magnetosphere 2D   the magnetosphere 3D

where

$$\alpha = S\hat{B}_\phi^2 + P\hat{B}_\theta^2$$

$$\gamma = 2(S - P)\hat{B}_\theta \hat{B}_\phi$$

$$\eta = S\left(1 + \frac{r}{R}\frac{\partial R}{\partial r}\right) + i\frac{D\hat{B}_\phi}{R}\frac{\partial R}{\partial \theta}$$

$$\lambda = \frac{S\hat{B}_\phi^2 + P\hat{B}_\theta^2}{r}\frac{1}{R}\frac{\partial R}{\partial \theta} - i\frac{D\hat{B}_\phi}{R}\frac{\partial R}{\partial r} \tag{23}$$

$$\mu = i\frac{D\hat{B}_\theta}{r}.$$

By solving this equation numerically we obtain the slow evolution of the
amplitude of the scalar potential. This allows us to reconstruct the potential

(phase and amplitude) Eq. (15). A plot of the equi-potential lines of the field in a poloidal plane of the torus is given in Fig. 8. In the figure it is possible to notice the distribution of the field that is maximum in a region around the equatorial plane of the antenna and propagate toward the magnetic axis where there is a focalization of the electromagnetic field around the caustic surface. In solving equations (22-23) great care has been taken in calculating the radial derivative of p_r (at constant θ): $\partial p_r/\partial r$ and the poloidal derivative of p_θ (at constant r): $\partial p_\theta/\partial\theta$. The space surface p_r and p_θ were calculated previously by the ray tracing solver (see plot Figs. 6 and 7).

As an example of the propagation of a radio signal (whistler wave) in the ionosphere we simulate the signal emitted by Ω-navigation transmitter (≈ 10 kW) in North Dakota ($\theta_g = 46°22'$, $\phi_g = 98°20'$) in a frequency range of 10.2-13.1 kHz.[25] We simulate the propagation of the first pulse at $f = 10.2$ kHz with duration of 1.1 sec. Figure 9 shows, in a meridian plane of a spherical geometry, a boundles of rays starting from the dipole antenna used in the experiment. It is possible to recognize in the plot the Earth (in scaled dimensions), while the rays starting at the antenna with $p_\theta = 0$ and $p_\phi = 0$, and p_r (calculated from the dispersion relation):

$$p_r = -\frac{[D^2 - S(P+S)]\hat{B}_\theta^2 - 2PS\hat{B}_r^2}{2[S\hat{B}_\theta^4 + (P+S)\hat{B}_\theta^2\hat{B}_r^2 + P\hat{B}_r^4]}$$

$$\pm\sqrt{\left[\frac{[D^2 - S(P+S)]\hat{B}_\theta^2 - 2PS\hat{B}_r^2}{2[S\hat{B}_\theta^4 + (P+S)\hat{B}_\theta^2\hat{B}_r^2 + P\hat{B}_r^4]}\right]^2 - \frac{P(S^2 - D^2)}{S\hat{B}_\theta^4 + (P+S)\hat{B}_\theta^2\hat{B}_r^2 + P\hat{B}_r^4}}$$

$$(24)$$

propagate straightway with diverging trajectories. The equiphase surfaces are also depicted on the same plot. For this run we have used a parabolic dependence of the density with the ionosphere height from $R_{iniz} = 100$Km to $R_{fin} = 500$Km. We have also included in the density model the fact that the ionization is field-aligned.[26] Considering a simple Earth magnetic dipole with radial $B_r(r,\theta) = -2B_0(R_{Earth}/r)^3\cos\theta$ and polar $B_\theta(r,\theta) = -B_0(R_{Earth}/r)^3\sin\theta$ components ($B_\phi = 0$), the equation for a magnetic field line is $R_{eq} = (r/\sin^2\theta)$, where $R_{eq} = R_{Earth} = 6370$Km is the Earth radius on the equatorial plane. The density profile is then depending from r and θ, we have

$$n(r,\theta) = n_0\left(1 + C_a\exp\left[-\left(\frac{b(\hat{r},\theta) - b_0}{\Delta}\right)^2\right]\right)[1 - (C_e\hat{r})^2]^{\alpha_d}, \quad (25)$$

where C_a, C_e, b_0, n_0, α_d and Δ are constants fixed by the experimental measure of the ionosphere density profile. Moreover, $\hat{r} = r/(R_{Earth} + R_{iniz})$, while $b = \hat{r}/\sin^2\theta$. In Fig. 10 a 3D plot of the rays is shown in spherical coordinate. In this case the rays are crossing the ionosphere layer with weak dispersion and traveling without radial reflection. The electric field polarization pertain to that of the launched whistler and no conversion in the short wavelength quasi-electrostatic LHW is absorbed.

5. Conclusions

We have presented an asymptotic analysis of the Maxwell equation which describes the propagation of the electromagnetic pulses in an ionized medium like that of laboratory plasmas (trapped in "tokamak devices" for studies oriented to the thermonuclear fusion research) and/or that of the propagation of radio signal in the terrestrial ionosphere. This analysis is based on the WKB expansion of the electric field in terms of a rapid varying phase and amplitude which varies on a much slower scale. The difficulty of the full solution of the coupled system of the PDE Maxwell equations is overcome by reducing the problem to a much simpler integration of the ray trajectories (system of ODE) at the lowest order and at the next order to the solution of the transport equation for the electric field amplitude. Example of the propagation of the Lower Hybrid wave in a "tokamak plasma" excited by external antenna structure is presented and discussed. A short discussion of the propagation of an electronic whistler propagating in the upper ionosphere is also performed.

References

1. N. Froman and P. O. Froman, *J.W.K.B. approximation*. North Holland Publ. Co., Amsterdam (1965).
2. R. Gans, *Ann. Physik* **47**, 709 (1915).
3. H.A. Kramers, *Z. Physik* **39**, 828 (1926).
4. H. Jeffreys, *Proc. London Math. Soc.* **23**, 428 (1924).
5. L. Brillouin, *Comptes Rendus de l'Academie des Sciences* **183**, 24 (1926).
6. G. Wentzel, *Z. Physik* **38**, 518 (1926).
7. K.G. Budden, *Radio Waves in the Ionospere* . Cambridge University Press, Cambridge (1961).
8. T.H. Stix, *Waves in Plasmas*. Springer Science & Business Media, New York (1992).
9. V.L. Ginzburg, *Propagation of Electromagnetic Waves in Plasma*. Gordon & Breach Science Publishers Ltd (1962).

10. M. Brambilla, *Kinetic Theory of Plasma Waves.* Clarendon Press, Oxford (1998).
11. P.C. Clemmow and J. Heading, *Proc. Cambridge Phil. Soc.* **50**, 319 (1954).
12. K.G. Budden and P.C. Clemmow, *Proc. Cambridge Phil. Soc.* **53**, 669 (1957).
13. M. Born and E. Wolf, *Principles of Optics: Electromagnetic Theory of Propagation, Interference and Diffraction of Light.* Pergamon Press, Oxford (1959).
14. J.R. Wait, *Radio Sci. J. Res.* **68D**, 1187 (1964).
15. S. Weinberg, *Phys. Rev.* **126**, 1899 (1962).
16. I. Bernstein, *Phys. Fluids* **18**, 320 (1975).
17. L. Friedland and I. Bernstein, *IEEE Trans. On Plasma Sci.*, **PS-8**, 90 (1980).
18. Y. Baranov and I. Fedorov, *Soviet Phys. Tech. Phys. Lett.* **4**, 322 (1978); Y. Baranov and I. Fedorov, *Nucl. Fusion* **20**, 1111 (1980).
19. M. Brambilla and A. Cardinali, *Plasma Phys.* **24**, 1187 (1982).
20. M. Bornatici, R. Cano, O. De Barbieri and F. Engelmann, *Nucl. Fusion* **23**, 1153 (1983).
21. E. Camporeale, G. L. Delzanno, P. Colestock, *J. Geophysical Res.: Space Physics* **117**, A10 (2012).
22. I.B. Bernstein, *Phys. Rev.* **109**, 10 (1958).
23. D.R. Smith, *J. Differential Equations* **68**, 383 (1987).
24. R. Spigler, M. Vianello, LiouvilleGreen, *Asymptotic Analysis* **48**, 267 (2006).
25. T.F. Bell, H.G. James, U.S. Inan, J.P. Katsufrakis, *J. of Geophysical Res.* **88A6**, 4813 (1983).
26. I. Yabroff, *J. Res. National Bureau of Standards D* **65**, 485 (1961).

Chapter 3

Simple exact and asymptotic solutions of the 1D run-up problem over a slowly varying (quasiplanar) bottom

S.Yu. Dobrokhotov, D.S. Minenkov, V.E. Nazaikinskii

A. Ishlinsky Institute for Problem in Mechanics of the Russian Academy of Sciences; Institute of Physics and Technology (Moscow, Russia)

B. Tirozzi

Dipartimento di Fisica, "Sapienza" University of Rome (Rome, Italy) and Enea Research Centre Frascati

We discuss localized asymptotic solution of the one-dimensional nonlinear shallow water equations over nonplanar bottom with depth vanishing at the origin (the "nonlinear run-up problem"). We use the semiclassical approximation and a generalization of the Carrier–Greenspan transformation to obtain a simple efficient formulas for the run-up of localized waves on the shore and the uprush of such waves.

1. Statement of the problem

The problem on the run-up of waves on a sloping beach is one important problem in the theory of tsunami waves. In the one-dimensional case, the mathematical statement of the problem is as follows: find the free surface elevation $\eta(x,t)$ and the velocity $u(x,t)$ from the Cauchy problem for the nonlinear shallow water equations

$$\eta_t + (D(x)u + \eta u)_x = 0, \qquad u_t + \left(\eta + \frac{1}{2}u^2\right)_x = 0 \qquad (1)$$

over a bottom of variable with the initial data

$$\eta|_{t=0} = \eta^0(x) \equiv \nu V\left(\frac{x-a}{\mu}\right), \qquad u|_{t=0} = 0 \qquad (2)$$

localized in a neighborhood of some point $x = a$, $a > 0$. Here x is the spatial coordinate, t is time, and the function $D(x)$ describing the bottom depth is assumed to be smooth. The point $x = 0$ corresponds to the shoreline (for zero free surface elevation), so that $D(x) > 0$ for $x > 0$ and $D(0) = 0$. Let

$$\gamma^2 = \left. \frac{\partial D(x)}{\partial x} \right|_{x=0}$$

be the bottom slope at the boundary of the basin. We consider the case in which $\gamma > 0$. As to the initial data, we assume that $V(\xi)$ is a smooth function that is compactly supported or at least rapidly decays as $|\xi| \to \infty$. The parameters $\mu \ll a$ and $\nu \ll D(a)$ characterize the wave length and amplitude at the initial time. Without loss of generality, we can assume that $a = 1$ and $\gamma = 1$. (This can always be achieved by an appropriate rescaling.) Then the wave length and height are small, $\nu \ll \mu \ll 1$. We seek a solution of Eqs. (1) on a half-line $[x_0(t), \infty)$, where the point $x_0(t)$ is determined from the condition

$$\eta(x,t) + D(x) \geq 0 \quad \text{for } x \geq x_0(t), \qquad \eta(x_0(t), t) + D(x_0(t)) = 0.$$

The point $x = x_0(t)$ is naturally referred to as the *shoreline position*, and $\eta^* = \max_t \eta(x_0(t), t)$ is called the *maximum run-up*.

It depends on the initial conditions whether there exists a global solution of the nonlinear problem (1), (2). In particular, the global solution exists if the parameter ν is sufficiently small; the solution has the form of a wave approaching the shore and then reflected by it. With increasing ν, the run-up wave front becomes steeper, and eventually the wave breaks. Needless to say, system (1) describes the wave propagation rather approximately and should not be used after wave breaking. Nevertheless, even this model gives some useful information about the the wave run-up and is often in good agreement with experimental data. In applications to tsunami problems, one is mainly interested in the free surface elevation $\eta(x,t)$, the shoreline position $x_0(t)$, and the threshold value of the initial wave amplitude at which the breaking process is triggered.

Now let us discuss the nonlinear approach to the description of waves arriving at the beach.[3] One wishes to have a theory that permits computing the increase in the wave height as the waves (in particular, tsunami waves) arrive at the beach. The problem is highly nonlinear and even the shallow water equation might prove inadequate for this situation. Most of the existing formulas pertain to the case of a planar bottom (e.g., see Refs. 13, 21,22). Our formulas, in a sense, generalize this approach to the case of a nonplanar bottom; we also present some rather simple solution formulas for special sources.

Let us outline the structure of the paper.

In Sec. 2, we recall the transformation introduced in Ref. 3. Section 3 presents some examples of exact solutions of the linearized system over a planar sloping bottom, and in Sec. 4 we introduce a bottom straightening transformation for the linearized system. The nonlinear shallow water equations for the case of a slowly varying nonplanar bottom are considered in Sec. 5. Section 6 provides the asymptotics of the solution of the linear run-up problem. Finally, some simple formulas for the shoreline position ("uprush") are given in Sec. 7.

2. Planar sloping bottom. Carrier–Greenspan transformation

First, consider the case in which $D(x) = x$ (i.e., the case of a planar bottom). In this case, there is a remarkably simple transformation due to Carrier and Greenspan,[3] which permits one to express solutions of the nonlinear system

$$\eta_t + (x u + \eta u)_x = 0, \qquad u_t + \left(\eta + \frac{1}{2} u^2\right)_x = 0 \qquad (3)$$

via solutions of the linear system

$$N_\tau + (z U)_z = 0, \qquad U_\tau + N_z = 0, \qquad (4)$$

which is obtained from (3) simply by *dropping the nonlinear terms*. (In the linear system, we have also changed the notation of the variables as follows: $t \to \tau$, $x \to z$, $\eta \to N$, and $u \to U$.)

We present this transformation in the following form[12] (see also Ref. 21). Let $(N(\tau, z), U(\tau, z))$ be a solution of the linear system (4) in some domain $\Omega_{(z,\tau)}$ of the variables (z, τ), and let $\Omega_{(x,t)}$ be the image of $\Omega_{(z,\tau)}$ under the mapping

$$x = z - N(z, \tau) + \frac{1}{2} U^2(z, \tau), \qquad t = \tau + U(z, \tau). \qquad (5)$$

Assume that the Jacobian

$$J_0 = \det \frac{\partial(x, t)}{\partial(z, \tau)} = 1 + U_\tau - N_z - U_\tau N_z + U U_z + N_\tau U_z$$

does not vanish anywhere in $\Omega_{(z,\tau)}$. Then the mapping (5) specifies a regular change of variables in $\Omega_{(z,\tau)}$ by the implicit function theorem,[a]

---

[a] We of course assume that $\Omega_{(z,\tau)}$ is connected and simply connected.

and we use this change of variables to define functions $\eta(x,t)$ and $u(x,t)$, $(x,t) \in \Omega_{(x,t)}$, by the formulas

$$\eta = N(z,\tau) - \frac{1}{2}U^2(z,\tau), \qquad u = U(z,\tau). \qquad (6)$$

Proposition 1. *The functions $\eta(x,t)$ and $u(x,t)$ parametrically defined by (5), (6) are a solution of system (3) in $\Omega_{(x,t)}$.*

Remark 1. 1. One can also express the solution (N,U) via (η, u) by using the change of variables[12]

$$z = x + \eta(x,t), \qquad\qquad \tau = t - u(x,t),$$
$$N(z,\tau) = \eta(x,t) + \frac{1}{2}u^2(x,t), \quad U(z,\tau) = u(x,t). \qquad (7)$$

2. The Jacobian J_0 does not vanish for sufficiently small amplitudes of the solution (N,U), and in this case systems (3) and (4) are equivalent and are related by the changes of variables (5)–(6) and (7).

3. The Carrier–Greenspan transformation (7) takes system (3) in the domain $\Omega_{(x,t)} = \{\gamma x + \eta(x,t) \geq 0\}$ with moving left boundary $x_0(t)$ to system (4) in the domain $\Omega_{(z,\tau)} = \{z \geq 0\}$ with fixed boundary $z_0(t) = 0$, because $z = x + \eta(x,t)$. Hence system (4) should be considered on the half-line $z \geq 0$ of the variable z.

4. The nonvanishing of the Jacobian means that the wave is reflected from the shore without breaking, which is often the case for tsunami waves.[21,22] Clearly, the relation $J_0 = 0$ can be used as an equation for the values of the wave parameters under which the wave breaking occurs.

Proof. Proposition 1 can be proved by various methods. The simplest proof is by straightforward differentiation with the subsequent substitution into the original equations. The derivation in Ref. 3 is based on the method of characteristics. One can also use group analysis methods, which permit constructing various transformations relating the solutions of the shallow water equations with horizontal and sloping bottom.[4,7]   □

System (4) can be rewritten in the form of two scalar equations for N and U,

$$N_{\tau\tau} - (zN_z)_z = 0, \qquad U_{\tau\tau} - (zU)_{zz} = 0. \qquad (8)$$

Recall that this system should be considered for $z \geq 0$. We see that the system is hyperbolic but degenerates on the boundary $z = 0$. This leads to some peculiarities in the statement of boundary conditions (see the next section and also Sec. 6).

3. Linear system over a planar sloping bottom. Some exact solutions

Let us present some simple exact solutions of the linear system (4) over the planar bottom $D(z) = z$.[4,7,12] Consider this system on the half-line $z \geq 0$ of the spatial variable z with the initial conditions

$$N_{\tau=0} = N^0(z), \quad U_{\tau=0} = U^0(z), \quad z > 0, \tag{9}$$

at $\tau = 0$. For the Cauchy problem (4), (9) to be well posed, one should impose an additional condition as $z \to 0$, because the wave propagation velocity degenerates on the line $z = 0$ (which can be viewed as a caustic of special type[8,12]). Namely, one should seek the solution in the class of functions with finite energy;[2,20] i.e., one should require that $\sqrt{z}N_z \in L^2(\mathbf{R}_+)$ for all τ. (This is just a special case of the general condition[8,25] for the wave equation with degeneration of this kind on the boundary.)

The wave equations (8) can be reduced to ordinary differential equations by the Hankel transform, which permits constructing an ample supply of exact solutions, including closed-form particular solutions in the form of Bessel functions of some complicated argument as well as more general solutions in integral form based on the inverse Hankel transform. Numerous oscillating and localized solutions of this kind were obtained and studied.[3,5,8,12,21–25] Note that, from the viewpoint of the semiclassical approximation, the Hankel transform occurring here is related to a classical canonical transformation associated with the equation $N_{\tau\tau} - (D(z)N_z)_z = 0$ with degenerating function $D(z)$. We return to this topic in Sec. 6.

The following family of simple exact solutions of the linear system (4) in the form of algebraic functions was obtained in Ref. 12 (they can also be obtained by group analysis methods[4,7]):

$$N(z,\tau) = \operatorname{Re} \frac{A(\tau + ib)}{(z - (\tau + ib)^2/4)^{3/2}},$$
$$U(z,\tau) = 2\operatorname{Re} \frac{A}{(z - (\tau + ib)^2/4)^{3/2}}, \tag{10}$$

where

$$A = \nu\mu^{3/2}(1+i)/(2\sqrt{a}), \quad b = -\beta\mu/\sqrt{a} + 2i\sqrt{a}, \quad \nu, \beta, \mu, a > 0.$$

The positive parameters $\nu, \beta, \mu, a > 0$ have the following meaning: as $\mu \to 0$, the functions $N(z,\tau)$ and $U(z,\tau)$ are localized at $\tau = 0$ in a neighborhood of the point $z = a$ (i.e., a determines the initial position of the wave and μ determines the wave length), ν determines the wave amplitude, and β determines the wave profile steepness (see Fig. 1).

A straightforward differentiation shows that (10) is indeed a solution of system (4).

The solution (10) describes a "solitary" wave that first runs from right to left, its maximum attaining the point $z = 0$ at time $\tau = 2\sqrt{a}$; at this time, the wave is reflected from the point $z = 0$ and then runs from left to right. The incident wave is bell-shaped and tends to the Dirac delta function as $\mu \to 0$ and $\nu \to \infty$, while the reflected wave has the shape of an "N-"wave (this effect was discovered in Refs. 21,22) and tends in the same limit to the Sokhotskii function $1/(y+i0)$. This phenomenon is well known in the theory of hyperbolic equations and systems as the metamorphose of the profile (or of the singularity) caused by passage through a focal point, a jump in the Maslov index, and the resulting application of the Hilbert transform to the original wave shape. For the one-dimensional wave equation, there is no metamorphose if the velocity is bounded away from zero, and in our case the metamorphose only occurs because $c^2(z) = z = 0$ for $z = 0$.

An application of the Carrier–Greenspan transformation to this solution under the assumption that the wave does not break (the corresponding Jacobian does not vanish) gives a solution of the nonlinear system. In this example, it is interesting that the effect is described by closed-form expressions, which permits analyzing and visualizing the solutions, say, with the help of *Wolfram Mathematica*.

Note also that the differentiation (as well as integration) of a solution $(N(z, \tau), U(z, \tau))$ with respect to time τ gives a new solution of system (4). Thus, if $(N(z, \tau), U(z, \tau))$ is a solution of the form (10) and $\mathcal{P}(\xi)$ is a polynomial in ξ and ξ^{-1}, then $(\mathcal{P}(\partial/\partial\tau)N(z, \tau), \mathcal{P}(\partial/\partial\tau)U(z, \tau))$ is a solution of system (4) as well. For $\tau = 0$, this new solution may have several distinct oscillations or coarsely model a step-like function.

4. Linear system over a variable bottom. "Bottom straightening" transformation

Now consider the linear (linearized) system with variable bottom given by a depth function $D(x)$ such that $D(x) > 0$ for $x > 0$ and $D(x) = \gamma^2 x f(x)$, $f(x) > 0$, in a neighborhood of the point $x = 0$. Thus, the system has the form

$$\eta_t + (D(x)u)_x = 0, \quad u_t + \eta_x = 0. \tag{11}$$

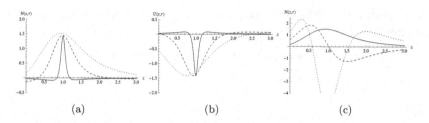

Fig. 1. Exact solutions: (a) and (b), $N(z, \tau)$ and $U(z, \tau)$ for $\beta = \nu = a = 1$, $\tau = 0$, and $\mu = 0.1$ (solid lines), 0.5 (dashed lines), and 1 (dotted lines); (c), $N(z, \tau)$ (solid lines), $N_\tau(z, \tau)$ (dashed lines), and $N_{\tau\tau}(z, \tau)$ (dotted lines) for $\beta = \nu = \mu = a = 1$ and $\tau = 0$.

By analogy with the preceding, for η and u we can write the scalar wave equations

$$\eta_{tt} - (D(x)\eta_x)_x = 0, \quad u_{tt} - (D(x)u)_{xx} = 0. \tag{12}$$

In this system, we pass from the independent variable x to a variable z chosen from the condition that the bottom becomes "planar" in the new variables. The change of variables has the form

$$z = Z(x), \quad Z(x) = \text{sign}(D(x))\left(\frac{1}{2}\int_0^x \frac{d\xi}{\sqrt{|D(\xi)|}}\right)^2. \tag{13}$$

The Jacobian

$$\tilde{J}(x) = \frac{dZ}{dx} = \sqrt{\frac{Z(x)}{D(x)}}$$

of this transformation is neither zero nor infinity for all $x \geq 0$ and hence (13) is a smooth one-to-one mapping.

The Jacobian $J(z) = \frac{d\xi}{dz}$ of the inverse change of variables $x = \xi(z)$ has the form

$$J(z) = \sqrt{\frac{D(\xi(z))}{z}},$$

or, equivalently,

$$z = J^{-2}(z)D(\xi(z)).$$

We proceed from (η, u) to the new unknowns (q, w) by the formulas

$$\eta = J^{-\frac{1}{2}}(z)q(z, t), \quad u = J^{-\frac{3}{2}}(z)w(z, t). \tag{14}$$

A straightforward differentiation shows that then system (11) and Eqs. (12) acquire the form

$$q_t + (wz)_z + za(z)w = 0, \qquad w_t + q_z - a(z)q = 0,$$
$$q_{tt} - (zq_z)_z + (za^2 + (za)_z)q = 0, \qquad w_{tt} - (zw)_{zz} + z(a^2 - a_z)w = 0,$$

$$(15)$$

where

$$a(z) = \frac{J_z}{2J}.$$

When constructing rapidly varying asymptotic solutions, the terms that contain a but do not contain the z-derivatives of q and w only produce small contributions, and hence the linear problem with arbitrary smooth variable bottom (velocity) described by the depth function $D(x) = \gamma^2 x f(x)$, $f(x) > 0$, can be reduced by formulas (14) to system (11) with $D(x) = x$. In the definitive formulas, one should express the variable z via the variable x by formula (13) and use the formula

$$J(Z(x)) = \tilde{J}^{-1}(x) = \sqrt{\frac{D(x)}{Z(x)}} \qquad (16)$$

for the Jacobian J, which is usually easier (and more efficient) in specific problems than to pass from the variable x to the variable z.

Take the solution (10) of Eqs. (15) and set

$$b = 2i\sqrt{z_0} + \mu b_0, \qquad A = \mu^{3/2} b_0^{3/2} e^{-i\pi/4} A_0,$$

where μ is a small parameter and $b_0, A_0, z_0 > 0$ are some constants. Accordingly, (10) becomes

$$q = \text{Re} \, \frac{A_0 e^{-i\pi/4}(t - 2\sqrt{z_0} + i\mu b_0)}{\left(\frac{z - (t - 2\sqrt{z_0})^2/4}{\mu b_0} + \frac{\mu b_0}{4} - i\frac{t - 2\sqrt{z_0}}{2} \right)^{3/2}},$$

$$w = 2\,\text{Re} \, \frac{A_0 e^{-i\pi/4}}{\left(\frac{z - (t - 2\sqrt{z_0})^2/4}{\mu b_0} + \frac{\mu b_0}{4} - i\frac{t - 2\sqrt{z_0}}{2} \right)^{3/2}}.$$

$$(17)$$

For $\mu \ll 1$, this solution is localized in a neighborhood of the trajectory $y = (t - 2\sqrt{y_0})^2$. In particular, for $t = 0$ it is concentrated in a neighborhood of the point y_0. This solution describes a localized solitary wave of characteristic length μ; the wave first runs from right to left, then is reflected from the boundary $x = 0$, and then runs from left to right with the new N shaped profile. The change of variables $z = Z(x)$ only slightly distorts the wave amplitude but substantially affects the trajectory of the wave crest. Namely, the trajectory can be found from the Hamiltonian system

$$\dot{p} = -|p|D_x(x), \quad \dot{x} = \frac{p}{|p|}D(x), \quad p|_{t=0} = 1, \quad x|_{t=0} = z_0. \tag{18}$$

The equation for the x-component can be rewritten as

$$\dot{x} = D(x) \quad \text{for} \quad p > 0; \qquad \dot{x} = -D(x) \quad \text{for} \quad p < 0. \tag{19}$$

Example 1. Consider the case in which

$$D(x) = \gamma^2 \frac{x}{1+x}. \tag{20}$$

Then

$$Z(x) = \frac{(\sqrt{x}\sqrt{1+x} + \text{arcsinh}\sqrt{x})^2}{4\gamma^2},$$

$$J(Z(x)) = \frac{2\sqrt{x}\gamma^2}{\sqrt{1+x}(\sqrt{x}\sqrt{1+x} + \text{arcsinh}\sqrt{x})}. \tag{21}$$

We take the solution (17) of Eqs. (15) and define y and J in (14) by formulas (21), thus obtaining an asymptotic solution for the case in which the bottom profile is given by (20). An elementary analysis of the resulting formulas shows that, for sufficiently large x, the solution behaves practically in the same way as the running-to-the-left solution $f(x + t)$ of the wave equation $\eta_{tt} - \eta_{xx} = 0$ with constant profile. However, when approaching the point $x = 0$, the profile undergoes the same changes as that of the wave with the degenerating velocity $c = \gamma\sqrt{x}$.

5. Slowly varying (quasi-planar) bottom. Modified Carrier–Greenspan transformation

In this section, η and u stand for the solution of the nonlinear system (1) and should not be mixed with the functions denoted by the same letters in the preceding section.

Now consider the nonlinear shallow water equations (1) assuming that the bottom is quasi-planar,[21,23] namely,

$$D(x) = \gamma^2 x(1 + O(\varepsilon)), \tag{22}$$

where $\varepsilon \to 0$ is a small parameter. Further, assume that we deal with initial data (2) such that the small parameter μ is of the same order as ε.

Then the solution of the linearized system has the property that the kth derivatives of N and U are of the order of ε^{-k}; in other words,

$$\varepsilon N_\tau, \varepsilon N_z, \varepsilon^2 N_{\tau\tau}, \ldots = O(1). \tag{23}$$

An asymptotic solution of this problem was obtained in Ref. 16 with the use of the above-mentioned bottom straightening change of variables. (See also Ref. 1, where the boundary value problem was considered.) The recipe for constructing the asymptotics in this problem is as follows: the change of variables (13) reduces the system to a nonlinear system with linear bottom and a perturbation that is small (of the order of $O(\varepsilon^2)$) compared with $\varepsilon\eta_t, \varepsilon\eta_x, \ldots$. After that, the change of variables (7) reduces the system to a perturbed linear system with linear bottom, the nonlinear perturbation being small.

In the case of the quasi-planar bottom (22), one has

$$Z(x) = \frac{x}{\gamma^2}(1 + O(\varepsilon)), \qquad \tilde{J}(x) = \frac{1}{\gamma^2} + O(\varepsilon) \neq 0.$$

Note that $\tilde{J}_x(x) = O(\varepsilon)$, and so the resulting system is indeed a weak perturbation of the linear system (11). Moreover, for the asymptotic solutions satisfying (23), the perturbation term is $O(\varepsilon^2)$ and does not affect the leading term of the asymptotics, which is only possible because of the substitution $Z(x)$. Namely, the following assertion holds.[16]

Proposition 2. *Let* $(N(z,\tau), U(z,\tau))$ *be the solution (or the leading term of the asymptotics of the solution) of system* (11). *Next, assume that* $J_0 \neq 0$, $D(x) = \gamma^2 x + O(\varepsilon)$, *and* N *and* U *satisfy* (23). *Then the following formulas give an asymptotic solution* $(\eta(x,t), u(x,t))$ *of system* (1):

$$\eta(x,t) = \tilde{J}^{-2}(x)\, q(Z(x), t) + O(\varepsilon), \quad u(x,t) = \tilde{J}^{-1}(x)\, w(Z(x), t) + O(\varepsilon),$$

where $Z(x)$ *and* $\tilde{J}(x)$ *are defined above and* $q(y,t), w(y,t)$ *are given parametrically (with parameters* z, τ*) by*

$$q(y,t) = N(z,\tau) - \frac{1}{2}U(z,\tau)^2, \qquad w(y,t) = U(z,\tau),$$

$$y = z - N(z,\tau) + \frac{1}{2}U(z,\tau)^2, \qquad t = \tau + U(z,\tau).$$

An example of the exact solution (N, U) and corresponding asymptotics (η, u) is presented in Fig. 2.

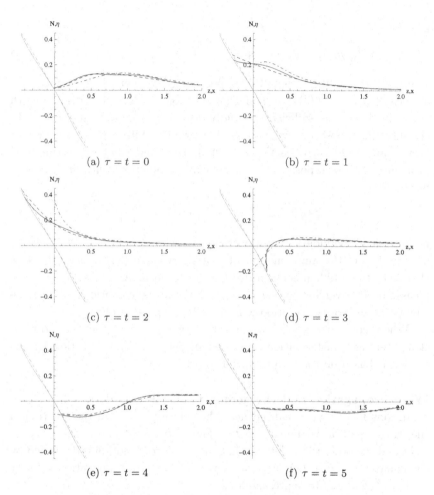

Fig. 2. Exact solution $N(z, \tau)$ (dot-dashed lines) and the corresponding asymptotics $\bar{\eta}(x, t)$ (dashed lines) for a planar bottom (dashed lines) and $\eta(x, t)$ (plain lines) for a quasi-planar bottom $D(x) = x(1 + 0.1 \sin(2\pi x))$ (gray solid line). The parameters are $\beta = \mu = a = 1$ and $\nu = 0.09$.

6. Asymptotics of the solution of the Cauchy problem for the linear system

Consider the linear system (11) with the initial conditions (2). To solve it, it suffices to solve the wave equation for η with appropriate initial conditions and then set $u(x, t) = -\int_0^t \eta_x(x, t)\, dt$. Thus, we deal with the Cauchy problem

$$\eta_{tt} - (D(x)\eta_x)_x = 0, \qquad \eta|_{t=0} = \nu V\left(\frac{x-a}{\mu}\right), \qquad \eta_t|_{t=0} = 0, \qquad (24)$$

where $x \in \mathbf{R}_+ = (0, \infty)$, ν and μ are positive parameters, $D(x)$ degenerates at zero ($D(0)=0$, $D'(0) > 0$, and $D(x) > 0$ for $x \in \mathbf{R}_+$), $V(y)$ is a smooth real-valued function sufficiently rapidly decaying at infinity, and $a > 0$. To simplify the exposition, we assume that $D(x)$ stabilizes to a constant for large x and $V(y)$ is compactly supported. By standard energy estimates,[2] problem (24) is well posed[8,20,25] in the class of functions with finite energy integral[b]

$$\mathcal{J}^2[\eta](t) = \frac{1}{2}\left(\|\eta_t\|^2 + \|D^{1/2}(x)\eta_x\|^2\right), \qquad (25)$$

no "classical" boundary condition being necessary at $x = 0$. We seek the asymptotics of the finite-energy solution of problem (24) as $\mu \to 0$. Note that since the problem is linear, we are free to normalize the solution as we please. It is convenient to take $\nu = \sqrt{\mu}$ in the initial conditions, so that the energy integral of the solution will be $O(1)$, $\mathcal{J}^2[\eta](t) = O(1)$.

When constructing the asymptotics, we will not use the bottom-straightening transformation described in Sec. 4. Our exposition in this section is based on the results in Refs. 8,11,17–19,25.

Wave propagation and the geometry of the problem. The initial conditions in (24) are localized near the point $x = a$, and the solution of the wave equation is the sum of two localized waves, one running to the right and the other to the left, so that the wave front motion is described by the equations $\dot{x} = \pm c(x)$, $x|_{t=0} = a$, where the wave propagation velocity $c(x) = \sqrt{D(x)}$ can be represented as

$$c(x) = \sqrt{x}\varkappa(x) \qquad (26)$$

with the smooth positive function $\varkappa(x) = \sqrt{D(x)/x}$ on $\overline{\mathbf{R}}_+ = [0, \infty)$. In finite time, the wave running to the left reaches the boundary $x = 0$, where it is reflected and then runs back to the right; this follows from the asymptotics given below. To construct the asymptotics, we need a description of this wave motion geometry in Hamiltonian terms. The Hamiltonian of the problem is $H(x,p) = c(x)|p|$, and the two running waves correspond to the trajectories $\Lambda_\pm$ of the Hamiltonian system

$$\dot{x} = c(x)\,\mathrm{sign}\,p, \qquad \dot{p} = -c'(x)|p| \qquad (27)$$

---

[b]From now on, $\|\cdot\|$ stands for the norm in $L^2(\mathbf{R}_+)$.

passing through the points $(x, p) = (a, \pm c(a)^{-1})$ (where the Hamiltonian is equal to 1). The trajectories lie in the space $T_0^* \mathbf{R}_+ = \{(x, p) : x > 0, \; p \neq 0\}$ and have the form shown in Fig. 3(a). The canonical transformation[c]

$$(x, p) \mapsto (q, E), \qquad q = 1/p, \quad E = xp^2, \tag{28}$$

maps $T_0^* \mathbf{R}_+$ into the open half-plane $\Phi = \{(q, E) : E > 0\}$. The inverse transformation, defined outside $\Phi_\infty = \{(q, E) \in \Phi : q = 0\}$, has the form

$$x = q^2 E, \qquad p = 1/q. \tag{29}$$

From now on we identify $T_0^* \mathbf{R}_+$ with $\Phi \setminus \Phi_\infty$ via (28) and (29). The change of variables (29) reduces the Hamiltonian $H(x, p)$ to the form $\mathcal{H}(q, E) = \varkappa(q^2 E)\sqrt{E}$, which extends by continuity to the entire Φ as a smooth function. The trajectories $\Lambda_\pm$ turn out to be parts of the single trajectory $\Lambda \subset \Phi$ determined by the equation $\mathcal{H}(q, E) = 1$ (see Fig. 3(b)).

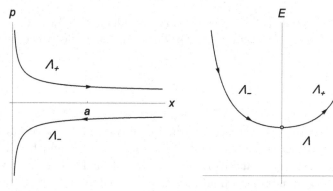

(a) The trajectories Λ_+ and Λ_- in the coordinates (x, p).

(b) In the coordinates (q, E), both trajectories Λ_+ and Λ_- prove to be parts of the same trajectory Λ.

Fig. 3.   The trajectories Λ_+, Λ_-, and Λ of the Hamiltonian system

We parametrize Λ with the time τ along the trajectories; if we adopt the convention that $\tau = 0$ at the point where $q = 0$, then Λ_+ is the region where $\tau > 0$, and Λ_- is the region where $\tau < 0$. Let us give a little more detailed coordinate description of Λ. The function

$$T(z) = 2 \int_0^z \frac{d\xi}{\varkappa(\xi^2)}$$

---

[c]That is, a transformation preserving the symplectic form, $dp \wedge dx = dE \wedge dq$.

has a smooth inverse function $z = z(T)$. Set $X(\tau) = z^2(\tau)$, $\tau \in (-\infty, \infty)$. The equations

$$x = X(\tau), \quad p = P(\tau) \equiv \frac{1}{z(\tau)\varkappa(z^2(\tau))}, \qquad \tau \neq 0, \tag{30}$$

parametrically define Λ_+ (for $\tau > 0$) and Λ_- (for $\tau < 0$) in the coordinates (x, p). On $\Lambda_\pm$, one can also express τ and p via x as follows:

$$p = p_\pm(x) \equiv \pm\frac{1}{c(x)}, \quad \tau = \tau_\pm(x) \equiv \pm\int_0^x \frac{d\xi}{c(\xi)} = \pm 2\int_0^{\sqrt{x}} \frac{d\xi}{\varkappa(\xi^2)}, \tag{31}$$

where the upper sign corresponds to Λ_+ and the lower, to Λ_-. The embedding $\Lambda \subset \Phi$ is described in the coordinates (q, E) by the parametric equations

$$q = Q(\tau) \equiv z(\tau)\varkappa(z^2(\tau)), \quad E = \mathcal{E}(\tau) \equiv \frac{1}{\varkappa^2(z^2(\tau))}, \quad \tau \in (-\infty, \infty). \tag{32}$$

For sufficiently small $\tau_0 > 0$, the function $q = Q(\tau)$, $\tau \in \Lambda_0 \equiv (-\tau_0, \tau_0)$ has a smooth inverse function $\tau = \tau(q)$. Thus, Λ_0 can be described by the equation

$$E = E(q) \equiv \mathcal{E}(\tau(q)). \tag{33}$$

Summarizing, we see that Λ is covered by three open intervals $\Lambda_\pm$ and Λ_0 such that x can be taken as a coordinate on $\Lambda_\pm$ and q as a coordinate on Λ_0. The equations specifying $\Lambda_\pm$ and Λ_0 read $p = p_\pm(x)$ and $E = E(q)$, where the functions $p_\pm(x)$ and $E(q)$ are given by (31) and (33), respectively. The intervals $\Lambda_\pm$ with the coordinate x will be referred to as the *nonsingular canonical charts* on Λ, and Λ_0 with the coordinate q will be referred to as the *singular canonical chart* on Λ.

Note also the useful formula

$$d\tau = P(\tau)\,dX(\tau) = 2\mathcal{E}(\tau)\,dQ(\tau) + Q(\tau)\,d\mathcal{E}(\tau)$$
$$= \mathcal{E}(\tau)\,dQ(\tau) + d(Q(\tau)\mathcal{E}(\tau)). \tag{34}$$

Reduction of problem (24) to a problem with rapidly oscillating initial conditions. Let $\varphi_0(x)$ be a smooth function supported in a sufficiently small neighborhood of the point $x = a$, and let

$$S_+(x) = \tau_+(x) - \int_0^a \frac{d\xi}{c(\xi)}.$$

Then the initial function in (24) admits the representation[8,11]

$$\nu V\left(\frac{x-a}{\mu}\right) = 2\nu \operatorname{Re} \int_\mu^\infty \widetilde{V}(\rho) e^{\frac{i\rho c_0}{\mu} S(x)} \varphi_0(x)\, d\rho + R, \quad |||R||| = O(\mu),$$
(35)

where $\widetilde{V}(\rho) = (2\pi)^{-1} \int_{-\infty}^\infty e^{-i\rho y} V(y,\mu)\, dy$ is the Fourier transform of $V(y)$, $\nu = \sqrt{\mu}$, and

$$|||\psi||| = \left[\|\psi\|_{L_2(\mathbf{R}_+)}^2 + \|c(x)\psi_x\|_{L_2(\mathbf{R}_+)}^2 \right]^{1/2}.$$
(36)

We use this representation to reduce problem (24) to a problem with rapidly oscillating initial conditions. Namely, consider the Cauchy problem

$$h^2 u_{tt} - h^2 (c^2(x) u_x)_x = 0, \quad u|_{t=0} = e^{\frac{i}{h} S_+(x)} \varphi_0(x), \quad u_t|_{t=0} = 0. \quad (37)$$

Let $u_{as}(x,t,h)$ be an asymptotic solution of (37) modulo $O(h^2)$ as $h \to 0$; more precisely, assume that $u_{as}(x,t,h)$ satisfies the initial conditions in (37) exactly and, being substituted into the equation, produces a left-hand side that is $O(h^2)$ in the norm of $L_2(\mathbf{R}_+)$ locally uniformly with respect to t. Set

$$\eta_{as}(x,t,\mu) = 2\sqrt{\mu} \operatorname{Re} \int_\mu^\infty \widetilde{V}(\rho) u_{as}\left(x,t,\frac{\mu}{\rho c_0}\right) d\rho.$$
(38)

Standard energy estimates for problem (24) imply the following assertion.

Proposition 3. *The function* (38) *is the asymptotics of the exact solution* $\eta(x,t,\mu)$ *of the Cauchy problem* (24) *in the sense that*

$$|||\eta_{as} - \eta||| + \|(\eta_{as} - \eta)_t\| = O(\sqrt{\mu})$$

locally uniformly with respect to t.

Solution of problem (37). The solution of problem (37) is given by a generalization[12,17–19,25] of the canonical operator[14,15] on Λ. Let us present the corresponding construction.

Consider the nonsingular canonical charts $\Lambda_\pm$ of the first type (a *nonsingular chart*). In these charts, we define the action functions

$$S_\pm(x) = \tau_+(x) - \int_0^a \frac{d\xi}{c(\xi)}$$

and the local canonical operators

$$K_\pm^h \colon C_0^\infty(\Lambda_\pm) \longrightarrow L_2(\mathbf{R}_+)$$

with small parameter $h \to 0$ by the formula

$$[K_{\pm}^h \varphi](x) = e^{-i\pi \frac{m_{\pm}}{2}} e^{\frac{i}{h} S_{\pm}(x)} \varphi(\tau_{\pm}(x)) c^{-1/2}(x),$$

where the *Maslov index* $m_{\pm}$ of the nonsingular chart $\Lambda_{\pm}$ is given by

$$m_+ = 0, \qquad m_- = -1.$$

In the singular canonical chart Λ_0, we define the action function

$$S_0(q) = \tau(q) - \mathcal{E}(\tau(q))q - \int_0^a \frac{d\xi}{c(\xi)}$$

and the local canonical operator

$$K_0^h : C_0^\infty(\Lambda_0) \longrightarrow L_2(\mathbf{R}_+)$$

by formula (2.22) in Ref. 8 with the substitution $p = q^{-1}$,

$$[K_0^h \varphi](x) = \frac{e^{i\pi\left(\frac{1}{4} - \frac{m_0}{2}\right)}}{\sqrt{2\pi h}} \int_{-\infty}^{\infty} e^{\frac{i}{h}\left(S_0(q) + \frac{x}{q}\right)} \left[\varphi(\tau)|Q'(\tau)|^{-1/2}\right]\Big|_{\tau=\tau(q)} \frac{dq}{q},$$

where $m_0 = 1$ is the *Maslov index of the singular chart* Λ_0 and the integral is understood (for $x > 0$) as an oscillatory integral. Now we define the canonical operator

$$K_\Lambda^h : C_0^\infty(\Lambda) \longrightarrow L_2(\mathbf{R}_+)$$

on Λ by the formula

$$K_\Lambda^h \varphi = K_+^h(e_+\varphi) + K_-^h(e_-\varphi) + K_0^h(e_0\varphi),$$

where $e_+ + e_0 + e_- = 1$ is a smooth partition of unity on Λ subordinate to the cover $\Lambda_+ \cup \Lambda_0 \cup \Lambda_- = \Lambda$.

The initial condition in (37) can be represented in the form

$$e^{\frac{i}{h} S_+(x)} \varphi_0(x) = [K_\Lambda^h \varphi](x),$$

where the function $\varphi \in C_0^\infty(\Lambda)$ is given by the formula

$$\varphi(\tau) = \begin{cases} \varphi_0(X(\tau))c^{1/2}(X(\tau)), & \tau > 0, \\ 0, & \tau \le 0. \end{cases}$$

Accordingly, the asymptotic solution of the Cauchy problem (37) modulo $O(h^2)$ is given by (cf. Sec. 2.3 in Ref. 8)

$$u_{as}(x,t,h) = \frac{1}{2}\left(e^{-i\frac{t}{h}}[K_\Lambda^h \varphi(\tau - t)](x) + e^{i\frac{t}{h}}[K_\Lambda^h \varphi(\tau + t)](x)\right).$$

Solution of problem (24). To obtain the solution of the original problem, it remains to use formula (38). We obtain

$$\eta_{as}(x,t,\mu) = \sqrt{\mu}\,\mathrm{Re}\int_\mu^\infty \tilde{V}(\rho)$$

$$\times \left(e^{-i\frac{c_0 t\rho}{\mu}}[K_\Lambda^{\frac{c_0\rho}{\mu}}\,\varphi(\tau-t)](x) + e^{i\frac{c_0 t\rho}{\mu}}[K_\Lambda^{\frac{c_0\rho}{\mu}}\,\varphi(\tau+t)](x)\right)d\rho, \quad (39)$$

and further simplification of this formula (which will be presented elsewhere) provides fairly explicit expressions for the asymptotic solution of problem (24).

Note that a generalization of the results presented in Sec. 6 to the two-dimensional and higher-dimensional cases is given in Refs. 9,10,18,19.

7. Uprush

For a tsunami a very important question is how far on the beach will the wave propagate. We want to study the dependence of the uprush (i.e. the maximal distance $- \min_t x_0(t) = -x_0(t^*) = -x_0^*$) on the amplitude and the wave length. The time t^* of the maximal distance in system (1) corresponds to the time τ^* in the linear system (4) at which the elevation $N(0,\tau)$ at the point $z = 0$ is maximal:

$$\max_\tau N(0,\tau) = N(0,\tau^*) = N^*.$$

We consider the exact solution (10) with $\mu = \beta = 1$ and parameters $a \sim 1$ and $\nu \ll 1$ which corresponds to the wave length and the amplitude:

$$N(0,\tau) = -\frac{4\nu}{\sqrt{a}}\frac{(\tau - 2\sqrt{a})^2 - 1/a - 2(\tau - 2\sqrt{a})/\sqrt{a}}{((\tau - 2\sqrt{a})^2 + 1/a)^2}.$$

In this case the time τ^* can be found analytically. The derivative

$$N_\tau(0,\tau) = -\frac{4\nu}{\sqrt{a}}\frac{(2a - 1 - \tau\sqrt{a})(1 + 4a(2+a) - 4\sqrt{a}(1+a)\tau + a\tau^2)}{(1 + a(\tau - 2\sqrt{a})^2)^3} = 0$$

vanishes at $\tau = \frac{2a-1}{\sqrt{a}}$ and $\tau = \frac{2(a+1)\pm\sqrt{3}}{\sqrt{a}}$ and the maximum of $N(0,\tau)$ is attained at $\tau^* = \frac{2(a+1)+\sqrt{3}}{\sqrt{a}}$ and is equal to $N^* = \frac{\nu}{2}(5+3\sqrt{3})\sqrt{a}$. It is clear that at that time $U(0,\tau^*) = 0$.

Now we use Proposition 2 to obtain the corresponding time t^* in system (1), which is $t^* = \tau^* + U(0,\tau^*) = \tau^* = \frac{2(a+1)+\sqrt{3}}{\sqrt{a}}$. The maximum distance

x_0^* is found from the equation $Z(x_0^*) = -N(0, \tau^*) + \frac{1}{2}U(0, \tau^*)^2 = -N^*$. Since $Z(x) = \frac{x}{\gamma^2}(1 + O(\varepsilon))$, we see that the uprush is

$$-x^* = \gamma^2 \frac{\nu}{2}(5 + 3\sqrt{3})\sqrt{a}(1 + O(\varepsilon)).$$

Acknowledgements

The research was supported by the RITMARE program (CINFAI) and by RFBR grant 14-01-00521a.

References

1. M. Antuono and M. Brocchini, *Studies in Appl. Math.* **124**(1), 85–103 (2010).
2. M.Sh. Birman and M.Z. Solomyak, *The Spectral Theory of Self-Adjoint Operators in Hilbert Space*. Leningrad State University, Leningrad (1980) [in Russian].
3. G.F. Carrier and H.P. Greenspan, *J. Fluid Mech.* **4**(1), 97–109 (1958).
4. Yu.A. Chirkunov, S.Yu. Dobrokhotov, S.B. Medvedev and D.S. Minenkov, *Teor. Mat. Fiz.* **178**(3), 322–345 (2014); English transl.: *Theor. Math. Phys.* **178**(3), 278–298 (2014).
5. I. Didenkulova and E. Pelinovsky, *Nonlinearity* **24**, R1–R18 (2011).
6. *Digital Library of Mathematical Functions*, National Institut of Standards and Technology, http://dlmf.nist.gov/10.51.
7. S.Yu. Dobrokhotov, S.B. Medvedev and D.S. Minenkov, *Mat. Zametki* **93**(5), 716–727 (2013); English transl.: *Math. Notes* **93**(5), 704–714 (2013).
8. S.Yu. Dobrokhotov, V.E. Nazaikinskii and B. Tirozzi, *Russ. J. Math. Phys.* **17**(4), 434–447 (2010).
9. S.Yu. Dobrokhotov, V.E. Nazaikinskii and B. Tirozzi, *Algebra i Analiz* **22**(6), 67–90 (2010); English transl.: *St. Petersburg Math. J.* **22**(6), 895–911 (2011).
10. S.Yu. Dobrokhotov, V.E. Nazaikinskii and B. Tirozzi, *Russ. J. Math. Phys.* **20** (4), 389–401 (2013).
11. S.Yu. Dobrokhotov, S.O. Sinitsyn and B. Tirozzi, *Russ. J. Math. Phys.* **14**(1),28–56 (2007).
12. S.Yu. Dobrokhotov and B. Tirozzi, *Uspekhi Mat. Nauk* **65**(1), 185–186 (2010); English transl.: *Russ. Math. Surveys* **65**(1), 177–179 (2010).
13. V.V. Ivanov, N.A. Ponomareva and A.A. Kharlamov, *Izv. Akad. Nauk. SSSR, Ser. Fiz. Atmosfery i Okeana* **26**(6), 659–664 (1990) [in Russian].
14. V.P. Maslov, *Perturbation Theory and Asymptotic Methods*, Moscow State University, Moscow (1965) [in Russian].
15. V.P. Maslov and M.V. Fedoryuk, *Semiclassical Approximation for Quantum-Mechanical Equations*. Nauka, Moscow (1976) [in Russian].
16. D.S. Minenkov, *Mat. Zametki* **92**(5), 721–730 (2012); English transl.: *Math. Notes* **92**(5), 664–672 (2012).
17. V.E. Nazaikinskii, *Mat. Zametki* **92**(1), 153–156 (2012); English transl.: *Math. Notes* **92**(1–2), 144–148 (2012).

18. V.E. Nazaikinskii, *Russ. J. Math. Phys.* **21**(2), 289–290 (2014).
19. V.E. Nazaikinskii, *Mat. Zametki* **96**(2), 261–276 (2014); English transl.: *Math. Notes* **96**(1–2), 248–260 (2014).
20. O.A. Oleinik and E.V. Radkevich, *Mathematical analysis*, 1969, pp. 7–252. Akad. Nauk SSSR Vsesojuzn. Inst. Naucn. i Tehn. Informacii, Moscow, 1971 [in Russian].
21. E.N. Pelinovskii, *Hydrodynamics of the Tsunami Waves*. IPF RAN, Nizhnii Novgorod (1996) [in Russian].
22. E.N. Pelinovsky and R.Kh. Mazova, *Natural Hazards* **6**(3), 227–249 (1992).
23. J.J. Stoker, *Water Waves: The Mathematical Theory with Applications*. **4**, Interscience, London (1957).
24. C.E. Synolakis, *J. Fluid Mech.* **185**, 523–545 (1987).
25. T. Vukašinac and P. Zhevandrov, *Russ. J. Math. Phys.* **9**(3), 371–381 (2002).

Chapter 4

Universality for couplings correlation in mean field spin glasses

A. Galluzzi, D. Tantari

Dipartimento di Matematica, "Sapienza" University of Rome (Rome, Italy)

F. Guerra

Dipartimento di Fisica, "Sapienza" University of Rome; Istituto Nazionale di Fisica Nuclere (INFN) (Rome, Italy)

In this paper we investigate a simple and novel (with respect to the classical one of choosing the couplings as Gaussian distributed or symmetrically distributed on ±1 values) type of *universality* that holds in mean-field spin-glasses . Even if there is a (mild) degree of correlations among couplings, the variational principles of statistical mechanics return sharply the same expression of the quenched free energy of the standard Sherrington-Kirkpatrick (SK) model. For positive couplings correlation, the model is formally equivalent to the sum of a SK model and a vanishing ordered perturbation, while for negative correlations we use interpolating techniques to get the replica symmetric (RS) and the broken replica symmetry (RSB) expressions of the relative quenched free energy.

1. Introduction

Spin glasses are *heaven for mathematicians*,[29] as, beyond their complexity that acts as a general framework for physicists, their rigorous control easily escapes whenever variations (even mild) with respect to the paradigmatic Sherrington-Kirkpatrick model (SK)[19] are considered. Already regarding the latter, a two-decade continuative effort of several researchers has been needed (and culminated in the celebrated tour de force of Guerra[21] and Talagrand[30]) in order to properly tackle -and frame into a formal scaffold[13]- the picture gave by Parisi in early '80.[25]

Several significative variations on theme appeared along the years: for instance the Viana and Bray proposal,[31] of studying the mean-field spin-glass on a Erdos-Renyi graph, that still lacks a complete rigorous Parisi-like theory, or the extension toward bipartite interacting systems (that found a more easily formal route[8,27]).

Another feature that attracted researchers in the past has been the so-called *universality* property of the SK model. In a nutshell the main result, achieved with heavy techniques stemmed from probability theory and functional analysis (see e.g.[15]), states that the degree of freedom of choosing the quenched couplings among spins symmetrically distributed over ± 1 or extracted from i.i.d. $\mathcal{N}[0,1]$ does not alter Parisi scenario (and an extension to bipartite models has been paved in[22]). To highlight the fragility of these invariances, it may be worth stressing that such universality is immediately lost when trying to allow the same freedom of choice in the spins more than in the couplings: choosing Gaussian spins instead of ± 1 standard Ising variables -hence dealing with a Gaussian spin glass- results in a replica-symmetric phase only,[6,10] a completely different scenario with respect to SK. Moreover other statistical mechanics mean field models do not exhibit such a property: for example the retrieval capabilities of the Hopfield model[24] with dichotomic patterns, and its generalizations,[1–5,28] disappear in the Analog Hopfield Model,[7,12,23] where the quenched soft patterns are gaussians.

While probably the most coveted universality-property aims to be the validity of Parisi prescription beyond the mean-field scenario,[14,32] here (far from that result), we consider a modified universality criterion, (with weaker correlations) with respect to the extension proposed by Chen in[16] (where universality is broken due to the presence of an extra magnetic field accounting for ferromagnetic inputs).

The model we investigate here belongs to a class of correlated models that has been introduced in,[17] where the existence of their thermodynamic limits was proved. In the present paper, after a streamlined exposition of the main observables and their related statistical mechanics package (Section 2), we perform our analysis by showing -in Sec.3- that the replica symmetric scenario for this models does coincide with the standard SK replica symmetric picture, while -in Sec.4- we show that Parisi solution is the solution even for this model, hence that the fully broken scenario of the SK is indeed the correct picture holding in this context too.

2. The model and its related definitions

The model is introduced by the following

Definition 1. The Hamiltonian of the Correlated Mean-Field Spin-Glass (CSG) reads as

$$H_N(\sigma; J) = -\frac{1}{\sqrt{2N}} \sum_{i,j=1}^{N} J_{ij}\sigma_i\sigma_j, \tag{1}$$

where $\sigma_i \in \{\pm 1\}$ and $i \in (1, ..., N)$, while the $\{J_{ij}\}$ is a family of correlated Gaussian random couplings defined by

$$\mathbb{E}[J_{ij}] = 0 \qquad \mathbb{E}[J_{ij}J_{kl}] = \left(1 - \frac{c}{N^2}\right)\delta_{(i,j),(k,l)} + \frac{c}{N^2}, \tag{2}$$

where $c \in \mathbb{R}$ tunes the degree of correlation among couplings (and clearly returns the standard Sherrigton-Kirkpatrick model if we pick $c \equiv 0$).

Once defined the partition function $Z_N(\beta, J)$ as

$$Z_N(\beta, J) = \sum_{\sigma} e^{\frac{\beta}{\sqrt{2N}} \sum_{i,j} J_{ij}\sigma_i\sigma_j} \tag{3}$$

with the inverse temperature β, the Gibbs measure

$$G_N(\sigma; \beta, J) = \frac{1}{Z_N} \exp^{-\beta H_N(\sigma; J)}$$

and all the standard statistical mechanics package, as the average over the $\{J_{ij}\}$ denoted by $\mathbb{E}(.)$, we are interested in the explicit expression for the quenched average of the free energy density $f(\beta)$ (strictly speaking the pressure $\alpha(\beta)$), introduced as

Definition 2. The quenched free energy for the CSG is defined as

$$\alpha(\beta) = \lim_{N \to \infty} \alpha_N(\beta) = -\beta f(\beta) = \lim_{N \to \infty} N^{-1}\mathbb{E}\ln Z_N(\beta, J). \tag{4}$$

Note that we will use the subscript N when observable are evaluated at fixed N, while we will omit any subscript whenever referring to the thermodynamic limit.

Further, let us introduce some basic notation: for a function $f(\sigma, \mathbf{J})$ of the degrees of freedom σ and the coupling $\mathbf{J}$ we introduce states and averages as

Definition 3. We define the Boltzmann state ω as

$$\omega(f) = \frac{\sum_{\sigma} f(\sigma) e^{-\beta H(\sigma)}}{\sum_{\sigma} e^{-\beta H(\sigma)}}$$

and the product state $\Omega = \omega \times \cdots \times \omega$ as the generalized thermal average over different replicas of the system, further

$$\langle f \rangle = E\Omega(f)$$

is the average first on the thermal weight and then over the quenched couplings.

Most of the thermodynamic informations about the system is encoded in the covariance of the Hamiltonian, considered as a family (parametrized by σ) of Gaussian correlated random variables. By direct computation

$$\mathbb{E}[H_N(\sigma^1; J)H_N(\sigma^2; J)]$$

$$= \frac{1}{2N} \sum_{i,j=1}^{N} \sum_{k,l=1}^{N} \sigma_i^1 \sigma_j^1 \sigma_k^2 \sigma_l^2 \left(\left(1 - \frac{c}{N^2}\right) \delta_{(i,j),(k,l)} + \frac{c}{N^2} \right) \qquad (5)$$

$$= \frac{N}{2} q_{\sigma^1 \sigma^2}^2 + \frac{Nc}{2} m^2(\sigma^1) m^2(\sigma^2), \qquad (6)$$

where we implicitly introduced the following order parameters

Definition 4. We define the magnetization of the system as $m(\sigma) = \frac{1}{N} \sum_{i=1}^{N} \sigma_i$ and the two-replica overlap as $q_{\sigma^1 \sigma^2} = \frac{1}{N} \sum_i \sigma_i^1 \sigma_i^2$.

Remark 1. Note that, as a corollary of eq. (6), choosing as starting Hamiltonian eq. (1) or the more familiar $H_N(\sigma; J) = -1/\sqrt{N} \sum_{i<j}^{N} J_{ij}\sigma_i\sigma_j$ implies only a shift in the correlation $c \to 2c$ (just consider the random variable $(J_{ij} + J_{ji})/\sqrt{2}$). Note further that the scaling of the correlation with the system size (i.e. $\propto N^{-2}$) ensures formally, still through formula (6), the correct (linear in N) extensivity for the internal energy.

We will need to compare the quenched free energy of the CSG with the Sherrington-Kirkpatrick one, we report the expressions for the latter, both at the replica symmetric and broken replica levels in the following

Proposition 1. *The replica symmetric quenched free energy of the Sherrington-Kirkpatrick model reads as*

$$\alpha_{RS}^{SK}(\beta) = \ln 2 + \mathbb{E}_z \log \cosh(\beta\sqrt{\bar{q}}z) + \frac{\beta^2}{4}(1 - \bar{q})^2, \qquad (7)$$

where the overlap $\bar{q}$ obeys

$$\bar{q} = \mathbb{E}_z \tanh^2 \left(\beta\sqrt{\bar{q}}z \right), \qquad (8)$$

while the related broken replica quenched free energy reads as

$$\alpha_{RSB}^{SK}(\beta) = \inf_{x(q)} \left[\ln 2 + f(0,0;x,\beta) - \frac{\beta^2}{2} \int_0^1 qx(q)dq \right], \qquad (9)$$

where the infimum is taken with respect to all non-decreasing functions $x : [0,1] \ni q \to x(q) \in [0,1]$ *and* $f(q,y;x,\beta)$ *is the solution of the Parisi equation*

$$(\partial_q f)(q,y) + \frac{1}{2}(\partial_{yy} f + x(q)(\partial_y f)^2)(q,y) = 0, \qquad (10)$$

with initial condition $f(1,y) = \ln\cosh(\beta y)$.

The main result of this paper concerns the existence of a generalized universality in mean field spin glasses for weakly correlated quenched noise, as stated in the following main

Theorem 1. *The free energy of the CSG does coincide* $\forall c \in \mathbb{R}$ *with the free energy of the Sherrington-Kirkpatrick model of the Proposition 1.*

As soon as $c > 0$ the result of Theorem 1 follows straightforwardly. In fact, in this case, the family of correlated quenched couplings defined in (2) can be represented by a decomposition through uncorrelated random variables:

$$J_{ij} = \sqrt{1 - c/N^2}\eta_{ij} + \sqrt{c}/N\eta, \qquad (11)$$

where η and η_{ij} are all i.i.d. Gaussian random variables. In terms of this decomposition, the CSG Hamiltonian can be rewritten as the sum of an SK Hamiltonian and an ordered perturbation term, i.e.

$$H_N(\sigma; \boldsymbol{\eta}) = -\sqrt{\frac{1 - c/N^2}{2N}} \sum_{i,j}^N \eta_{ij}\sigma_i\sigma_j - \sqrt{\frac{cN}{2}}\eta m_N^2(\sigma). \qquad (12)$$

The last term (ferromagnetic or antiferromagnetic depending on the quenched realization of η) does not affect the free energy in the thermodynamic limit, i.e.

$$\alpha_N^{SK}(\beta) - \sqrt{\frac{c}{2N}}\mathbb{E}(|\eta|) \leq \alpha_N(\beta) \leq \alpha_N^{SK}(\beta) + \sqrt{\frac{c}{2N}}\mathbb{E}(|\eta|). \qquad (13)$$

On the contrary, if $c < 0$, the decomposition (11) no longer holds, the coupling correlation cannot be represented through a common ordered term but tends to emphasize in some sense the disorder (decreasing the probability that two different couplings have the same sign) and thus the frustration of the system. For this reason we will prove universality by direct computation of the free energy, in the replica symmetric (RS) and the replica symmetry breaking (RSB) scenario, showing that the same variational principles of Proposition 1 hold.

3. Correlation-Universality at the replica symmetric level

Hereafter we exploit the strategy to formulate our replica-symmetric sum rule. The idea is very simple and is to compare through interpolation techniques, the replica symmetric free energy of the CSG and the replica symmetric free energy of the Sherrington-Kirkpartrick model to conclude that, in the thermodynamic limit, the two expressions become identical.

In order to achieve our result of showing explicitly the whole procedure to conclude our results, the explicit route we follow is to merge the Guerra's interpolation scheme[20] (which manages the real *glassiness* of the system) with the log-constrained entropy technique performed a' la Coolen[18] (which disentangles the pseudo-ferromagnetic and pseudo-antiferromagnetic contributions induced by the correlations).

We start comparing the original CSG-model with a one-body system, whose effective fields properly mimic the statistics of the real fields (at least at the first orders): thus we need to introduce additional correlated random (one-body) couplings J_i such that

$$\mathbb{E}[J_i] = 0 \; ; \qquad \mathbb{E}[J_i J_l] = \left(\bar{q} - \frac{b}{N} \right) \delta_{i,l} + \frac{b}{N},$$

where $\bar{q}$ is a scalar parameter (which will be understood later on as the replica-symmetric value of the overlap) and b is the degree of field's correlation and will be linked with the expected correlation among spins on different sites. With these fields we can introduce the random one body interaction $K(\sigma) = \sum_{i=1}^{N} J_i \sigma_i$, whose covariances read as

$$\mathbb{E}[K(\sigma^1)K(\sigma^2)]$$
$$= \sum_{i,j=1}^{N} \sigma_i^1 \sigma_j^2 \left(\bar{q} - \frac{b}{N} \delta_{i,j} + \frac{b}{N} \right) = N\bar{q}q_{\sigma^1\sigma^2} + bNm(\sigma^1)m(\sigma^2),$$

and define the following interpolating partition function $Z_N(\beta, \mathbf{J}, t)$ and quenched free energy $\alpha_N(\beta, t)$:

Definition 5. The partition function $Z_N(\beta, \mathbf{J}, t)$ interpolating between the original system (recovered at $t = 1$) and a one body spin-glass plus a non disordered term (obtained at $t = 0$) reads as

$$Z_N(\beta, \mathbf{J}, t) = \sum_\sigma e^{\sqrt{t}\beta H_N(\sigma;J)} e^{\sqrt{1-t}\beta K(\sigma)+(1-t)\frac{N\beta^2 c}{4}\left(m^2(\sigma)-\frac{b}{c}\right)^2}, \qquad (14)$$

and its related quenched free energy reads as

$$\alpha_N(\beta, t) = \frac{1}{N}\mathbb{E}\ln Z_N(\beta, t). \qquad (15)$$

Obviously, for $t = 1$ the quenched free energy of the original model is recovered, while for $t = 0$ a more tractable system is handled. The strategy now is to built a sum rule among the system at $t = 1$ and the system at $t = 0$, as (using the operator "dot" for the derivative w.r.t. the parameter t) stated by the following

Proposition 2. *It is possible to express in a sum-rule the (replica symmetric) interpolating quenched free energy of the CSG as follows:*

$$\alpha(\beta) = \lim_{N\to\infty} \alpha_N(\beta, t = 1) = \lim_{N\to\infty} \int_0^1 \dot\alpha(\beta, t')dt' + \alpha(\beta, t = 0). \qquad (16)$$

To this task we need to evaluate the t-streaming of the quenched free energy and show that it is possible to express $\dot\alpha(t)$ in terms of the fluctuations[9,20] of the order parameters (where the t-dependence is confined), such that, neglecting the latter, immediately results in a bound for the replica-symmetric free energy (i.e. with order parameters not-fluctuating in the thermodynamic limit[11]).

Proposition 3. *The following streaming equation holds for the (replica symmetric) interpolating quenched free energy of the CSG:*

$$\dot\alpha(\beta, t) = \langle \frac{\beta^2}{2N}\left(\frac{N}{2}\left(1 - q^2_{\sigma^1\sigma^2}\right) + \frac{Nc}{2}\left(m^4(\sigma^1) - m^2(\sigma^1)m^2(\sigma^2)\right)\right) \qquad (17)$$

$$- \frac{\beta^2}{2N}\left(N\bar q\left(1 - q_{\sigma^1\sigma^2}\right) + bN\left(m^2(\sigma^1) - m(\sigma^1)m(\sigma^2)\right)\right)\rangle_t$$

$$= \frac{\beta^2}{4}(1 - \bar q)^2 - \frac{\beta^2}{4}\langle(q_{\sigma^1\sigma^2} - \bar q)^2\rangle_t - \frac{\beta^2 c}{4}\langle\left(m\left(\sigma^1\right)m\left(\sigma^2\right) - \frac{b}{c}\right)^2\rangle_t.$$

Note that the last line expresses the t-streaming of $\alpha(\beta, t)$ as a term t-independent (i.e. $(\beta^2/4)(1 - \bar q)^2$) and a t-function (the dependence by t is nested in the Boltzmannfacktor) which is nothing but the (positive defined) fluctuations of the two order parameters around their means: neglecting such a term we obtain the replica symmetric bound.

However, to get the sum rule, we still need to evaluate the starting point $\alpha(\beta, t = 0)$, and we use here the Coolen's way,[18] namely, calling $b/c \equiv M^2$ and $H(m, M) = \frac{\beta^2 c}{4} \left(m^2(\sigma) - M^2\right)^2$, we can write

$$\alpha(\beta, t = 0) \tag{18}$$

$$= \frac{1}{N} \mathbb{E} \ln \sum_\sigma e^{\beta \sum_{i=1}^N J_i \sigma_i} e^{\frac{N\beta^2 c}{4}\left(m^2(\sigma) - \frac{b}{c}\right)^2}$$

$$= \frac{1}{N} \mathbb{E} \ln \int_{-1}^1 dm \int_{-\infty}^\infty \frac{d\hat{m}}{\sqrt{2\pi N}} e^{NH(m,M)} e^{-iN\hat{m}m} \sum_\sigma e^{\beta \sum_{i=1}^N J_i \sigma_i} e^{i\hat{m} \sum_{i=1}^N \sigma_i}$$

$$= \frac{1}{N} \mathbb{E} \ln \int_{-1}^1 dm \int_{-\infty}^\infty \frac{d\hat{m}}{\sqrt{2\pi N}} e^{N(H(m,M) - i\hat{m}m)} e^{N\left(\frac{1}{N}\sum_{i=1}^N \log \cosh(\beta J_i + i\hat{m}) + \ln 2\right)}$$

$$= \frac{1}{N} \mathbb{E} \ln \int_{-1}^1 dm \int_{-\infty}^\infty \frac{d\hat{m}}{\sqrt{2\pi N}} e^{N(H(m,M) - i\hat{m}m)} e^{N(\mathbb{E}_z \log \cosh(\beta \sqrt{\bar{q}} z + i\hat{m}) + \ln 2)}.$$

$$= \text{Extr}_{m,\hat{m}} \left[\ln 2 + \mathbb{E}_z \log \cosh\left(\beta \sqrt{\bar{q}} z + i\hat{m}\right) + H(m, M) - i\hat{m}m\right],$$

when we have computed the entropic term (i.e. the number of configuration with fixed magnetization m) introducing the integral representation of a delta function $\delta_{m(\sigma),m} = \int_0^1 \frac{d\hat{m}}{\sqrt{2\pi N}} e^{iN\hat{m}(m(\sigma) - m)}$ and we used the saddle point technique in the last equality to extimate the thermodynamic limit. Finding the extrema w.r.t. $m, \hat{m}$ we need to solve the following system

$$\partial_m[H(m, M) - i\hat{m}m + \ln 2 + \mathbb{E}_z \ln \cosh(\beta \sqrt{\bar{q}} z + i\hat{m})] = 0$$
$$\Rightarrow -i\hat{m} + \partial_m H(m, M) = 0, \tag{19}$$
$$\partial_{\hat{m}}[H(m, M) - i\hat{m}m + \ln 2 + \mathbb{E}_z \ln \cosh(\beta \sqrt{\bar{q}} z + i\hat{m})] = 0$$
$$\Rightarrow -im + i\mathbb{E}_z \tanh(\beta \sqrt{\bar{q}} z + i\hat{m}) = 0 \tag{20}$$

which implies the following

Lemma 1. *The starting point (Cauchy condition) for the propagation along the streaming of the (replica symmetric) interpolating quenched free energy of the CSG reads as*

$$\alpha(\beta, t = 0)$$
$$= \ln 2 + H(m, M) + \mathbb{E}_z \ln \cosh\left(\beta \sqrt{\bar{q}} z + \partial_m H(m, M)\right) - m\partial_m H(m, M).$$
$$m = \mathbb{E}_z \tanh\left(\beta \sqrt{\bar{q}} z + \partial_m H(m, M)\right), \tag{21}$$

with

$$H(m, M) = \frac{\beta^2 c}{4} \left(m^2(\sigma) - M^2\right)^2. \tag{22}$$

Now we can use Proposition 2 and, neglecting the positive defined terms concerning the fluctuations of the order parameters, we get that $\alpha(\beta, t = 1) \leq \alpha(\beta, t = 0) + \frac{\beta^2}{4}(1 - \bar{q})^2$. If we take the infimum over the couple (q, M) we get the following

Theorem 2. *The following replica symmetric bound for the quenched free energy of the CSG holds:*

$$\alpha(\beta, t = 1) \leq \inf_{\bar{q}, M}[\ln 2 + H(m, M) - m\partial_m H(m, M) + \mathbb{E}_z \ln \cosh(\beta\sqrt{\bar{q}}$$
$$+ \partial_m H(m, M))], \tag{23}$$

where $H(m, M) = \frac{\beta^2 c}{4}\left(m^2 - M^2\right)^2$, and whose extremization implies

$$\bar{q} = \mathbb{E}_z \tanh^2\left(\beta\sqrt{\bar{q}}z + \partial_m H(m, M)\right), \tag{24}$$
$$0 = \partial_{M^2} H(m, M) - m\partial_{M^2}\partial_m H(m, M) - m\partial_m\partial_{M^2} H(m, M). \tag{25}$$

The last result allows to conclude that $\partial_{M^2} H(m, M) = 0$, i.e. $m^2 = M^2$, and thus $H(m, M) = 0$, by which we obtain the following first theorem regarding the generalized correlation-universality.

Theorem 3. *The replica symmetric free energy of the correlated mean field spin glass is*

$$\alpha(\beta) = \ln 2 + \mathbb{E}_z \log \cosh(\beta\sqrt{\bar{q}}z) + \frac{\beta^2}{4}(1 - \bar{q})^2, \tag{26}$$

where the order parameter $\bar{q}$ satisfies

$$\bar{q} = \mathbb{E}_z \tanh\left(\beta\sqrt{\bar{q}}z\right), \tag{27}$$

namely, it is the same (replica symmetric) free energy of the standard Sherrington-Kirkpatrick model.

4. Broken replica universality

In this section we extend the scenario outlined above to the broken replica case, namely including Parisi Theory.

The way to proceed is clear despite calculations become sometimes a bit cumbersome: We need to extend our interpolating structure, by introducing the Guerra' broken-replica scheme[21] and derive once more the free energy of the CSG. At the end, by a simple comparison between this broken replica expression and the Sherrington-Kirkpatrick one, as they turn out to be equal in the thermodynamic limit, we obtain the result.

To this task, let us introduce two sequences of real numbers $\{q_l\}_{l=1}^K$ and $\{b_l\}_{l=1}^K$ and an increasing in $[0,1]$ succession $\{m_l\}_{l=1}^K$ with $m_0 = 0$ and $m_{K+1} = 1^*$; further we need also additional random (one-body) Gaussian fields J_i^l (with $i \in (1, ..., N)$ and $l \in (1, ..., K)$), such that $\mathbb{E}[J_i] = 0$ and $\mathbb{E}[J_i^l J_j^n] = \delta_{l,n} \cdot (\delta_{i,j} q_l + \delta_{i \neq j} b_l/N))$. The latter allows to introduce $K_l(\sigma) = \sum_{i=1}^N J_i^l \sigma_i$ and define an interpolating partition function $Z_N(\beta, \mathbf{J}, t)$ as stated by the next

Definition 6. The interpolating partition function useful in the broken replica scenario is

$$Z_N(\beta, \mathbf{J}, t) = \sum_\sigma e^{\sqrt{t}\beta H_N(\sigma; J)} e^{\sqrt{1-t}\beta \sum_{l=1}^K \sum_{i=1}^N J_i^l \sigma_i}, \tag{28}$$

together with its usual recursive relation[21]

$$Z_K = Z_N(\beta, \mathbf{J}, t); \quad Z_{K-1}^{m_K} = \mathbb{E}_K(Z_K^{m_K}), \quad Z_0^{m_1} = \mathbb{E}_1(Z_1^{m_1}),$$

which allows to finally define the quenched free energy as

$$\alpha_N(\beta, t) = \frac{1}{N}\mathbb{E}\ln Z_0. \tag{29}$$

Of course $\lim_{N \to \infty} \alpha_N(\beta, t = 1) = \alpha(\beta)$.

The above recursive relations for the partition function's chains imply a non-standard definition of deformed Boltzmann states which is worth stressing, in particular, defining $f_l = Z_l^{m_l}/\mathbb{E}_l(Z_l^{m_l})$ we introduce them as

$$\omega_l(.) = \mathbb{E}_{l+1}...\mathbb{E}_k(f_{l+1}...f_K\omega(.)),$$
$$\Omega_l(.) = \omega_l(.) \otimes ... \otimes \omega_l(.),$$
$$\langle .\rangle = \mathbb{E}(f_1...f_l\Omega_l(.)).$$

It may also be useful (to speed up calculations) to write explicitly the expectation of the covariances, namely

$$C_{\sigma\tilde{\sigma}} = \mathbb{E}[H_N(\sigma; J)H_N(\tilde{\sigma}; J)] = \frac{q_{\sigma\tilde{\sigma}}^2}{2} + \frac{c}{2}m^2(\sigma)m^2(\tilde{\sigma}), \tag{30}$$

$$B_{\sigma\tilde{\sigma}}^l = \mathbb{E}[K^l(\sigma)K^l(\tilde{\sigma})] = q_l q_{\sigma\tilde{\sigma}} + b_l m(\sigma)m(\tilde{\sigma}). \tag{31}$$

With a long but straightforward calculation, it is then possible to obtain the t-streaming of the quenched free energy as

$$\dot{\alpha}_N(\beta, t) = \frac{\beta^2}{2}\langle C_{\sigma\sigma} - \sum_{l=1}^K B_{\sigma\sigma}^l\rangle - \frac{1}{2}\sum_{l=1}^K (m_{l+1} - m_l)\langle C_{\sigma\tilde{\sigma}} - \sum_{n=1}^l B_{\sigma\tilde{\sigma}}^n\rangle_{l,t}, \tag{32}$$

---

*K in this context may be though of as the step of replica symmetric breaking performed in the framework of the replica trick.[25,29]

hence

$$\dot{\alpha}_N(\beta, t) = \frac{\beta^2}{2} \langle \frac{1}{2} + \frac{c}{2} m^4(\sigma) - \sum_{l=1}^{K} q_l - (\sum_{l=1}^{K} b_l) m^2(\sigma) \rangle_{K,t}$$

$$- \frac{\beta^2}{2} \sum_{l=0}^{K} (m_{l+1} - m_l) \langle \frac{1}{2} q_{\sigma\tilde{\sigma}}^2 + \frac{c}{2} m^2(\sigma) m^2(\tilde{\sigma}) \rangle$$

$$- \langle (\sum_{n=1}^{l} q_l) q_{\sigma\tilde{\sigma}} - (\sum_{n=1}^{l} b_l) m(\sigma) m(\tilde{\sigma}) \rangle_{l,t}. \qquad (33)$$

If, for the sake of simplicity, we call $Q_l = \sum_{n=1}^{l} q_n$ and $B_l = \sum_{n=1}^{l} b_n$, and, in order to have squared fluctuations as in the previous section, we add and cut to the above streaming the terms $\pm \frac{\beta^2}{4} \sum_l (m_{l+1} - m_l) Q_l^2$ and $\pm \frac{\beta^2}{4c} \sum_l (m_{l+1} - m_l) B_l^2$ we can write (calling $\mathcal{M} \equiv (m_{l+1} - m_l)$ to save space)

$$\dot{\alpha}_N(\beta, t) =$$

$$- \frac{\beta^2}{4} \sum_{l=0}^{K} \mathcal{M} \langle (q_{\sigma\tilde{\sigma}} - Q_l)^2 \rangle_{l,t} - \frac{\beta^2 c}{4} \sum_{l=0}^{K} \mathcal{M} \langle \left(m(\sigma) m(\tilde{\sigma}) - \frac{B_l}{c} \right)^2 \rangle_{l,t}$$

$$+ \frac{\beta^2}{4} \sum_l \mathcal{M} Q_l^2 + \frac{\beta^2}{4c} \sum_l \mathcal{M} B_l^2 - \frac{\beta^2}{4} + \langle \frac{\beta^2 c}{4} m^4(\sigma) - \frac{\beta^2}{2} B^K m^2(\sigma) \rangle_{K,t}$$

$$= - \frac{\beta^2}{4} \sum_{l=0}^{K} \mathcal{M} \langle (q_{\sigma\tilde{\sigma}} - Q_l)^2 \rangle - \frac{\beta^2 c}{4} \sum_{l=0}^{K} \mathcal{M} \langle (m(\sigma) m(\tilde{\sigma}) - \frac{B_l}{c})^2 \rangle_{l,t}$$

$$+ \frac{\beta^2}{4} (1 - \sum_{l=1}^{K} \mathcal{M} Q_l^2) + \frac{\beta^2 c}{4} \langle m^4(\sigma) - 2 \frac{B^K}{c} m^2(\sigma) + \sum_{l=0}^{K} \mathcal{M} (\frac{B_l}{c})^2 \rangle_{K,t},$$

by which it is immediate to conclude that, in order to neglect the proper fluctuation-source in the sum rule (for each step of replica symmetry breaking[9]), a "standard" interpolating procedure is not enough and another function $H(m(\boldsymbol{\sigma}), \mathbf{B})$ must be added to tackle the pseudo-ferromagnetic and pseudo-antiferromagnetic contributions, namely

$$H(m(\boldsymbol{\sigma}), \mathbf{B}) = \frac{\beta^2 c}{4} \left[m^4(\sigma) - 2 \frac{B^K}{c} m^2(\sigma) + \sum_{l=0}^{K} (m_{l+1} - m_l)(\frac{B_l}{c})^2 \right]. \qquad (34)$$

Hence we need to extend the interpolation through $\tilde{Z}_N(\beta, \mathbf{J}, t)$, the latter being introduced by this

Definition 7. The extended interpolating partition function able to handle both the glassy and the (ferro and antiferro) magnetics contributions, generated by the correlated couplings of the CSG, is

$$\tilde{Z}_N(\beta, \mathbf{J}, t) = \sum_{\sigma} e^{\sqrt{t}\beta H_N(\sigma;\mathbf{J})} e^{\sqrt{1-t}\beta \sum_{l=1}^{K}\sum_{i=1}^{N} J_i^l \sigma_i} e^{(1-t)NH(m(\sigma),\mathbf{B})}, \quad (35)$$

by which we can finally write (extending the symbol tilde to the free energy related to $\tilde{Z}_N(\beta, \mathbf{J}, t)$ the next

Proposition 4. *The following streaming equation holds for the (broken replica) interpolating quenched free energy of the CSG:*

$$\frac{d\tilde{\alpha}(\beta, t)}{dt} = -\frac{\beta^2}{4}\left(1 - \sum_{l=1}^{K}(m_{l+1} - m_l)Q_l^2\right) \quad (36)$$

$$-\frac{\beta^2}{4}\sum_{l=0}^{K}(m_{l+1} - m_l)\langle(q_{\sigma,\tilde{\sigma}} - Q_l)^2\rangle$$

$$-\frac{\beta^2 c}{4}\sum_{l=0}^{K}(m_{l+1} - m_l)\langle(m(\sigma)m(\tilde{\sigma}) - \frac{B_l}{c})^2\rangle_{l,t}$$

and, of course, $\lim_{N\to\infty} \tilde{\alpha}(\beta, t = 1) = \alpha(\beta)$.

As we can see, with the same procedure exploited in the previous section,

$$\tilde{\alpha}(\beta, t = 1) = \tilde{\alpha}(\beta, t = 0) + \int_0^1 dt \frac{d\alpha(\beta, t)}{dt}$$

$$\leq \tilde{\alpha}(\beta, t = 0) - \frac{\beta^2}{4}\left(1 - \sum_{l=1}^{K}(m_{l+1} - m_l)Q_l^2\right) \quad (37)$$

and taking the infimum we can finally obtain the following

Theorem 4. *The next sum rule for the broken replica quenched free energy of the correlated mean field spin glass holds*

$$\tilde{\alpha}(\beta, t = 1) \leq \inf_{Q_l, B_l}\left\{\tilde{\alpha}(\beta, t = 0) - \frac{\beta^2}{4}\left(1 - \sum_{l=1}^{K}(m_{l+1} - m_l)Q_l^2\right)\right\}.$$

The interpolating function evaluated at $t = 0$ can be written as in the following

Proposition 5. *At $t = 0$ the interpolating function can be straightforwardly computed as*

$$\tilde{\alpha}(\beta, t = 0) = \ln 2 + H(m, \mathbf{B}) - m\partial_m H(m, \mathbf{B}) + f(0, \partial_m H(m, \mathbf{B}); x(Q), \beta)$$

where $f(q, y; x(Q, \beta))$ is the solution of the Parisi equation as in Theorem 1 and m satisfies the following self-consistent equation

$$m = \left\langle \tanh(\beta \sum_{l=1}^{K} J_l + \partial_m H(m, \mathbf{B})) \right\rangle_K \qquad (38)$$

with $\langle \mathcal{O}(\mathbf{J}) \rangle_K = \mathbb{E}[f_1 \dots f_K \mathcal{O}(\mathbf{J}]$ and $J_l = \mathcal{N}[0, q_l]$, with $\mathbb{E}[J_l J_n] = \delta_{l,n} q_l$.

Proof. We can repeat the same scheme used in the replica symmetric case if we rewrite the interpolating function using the Ruelle probability cascedes as

$$\tilde{\alpha}(\beta, t = 0) = \lim_{N \to \infty} \frac{1}{N} \mathbb{E} \log \sum_{\sigma} \sum_{\alpha \in \mathbb{N}^K} \nu_\alpha e^{\beta \sum_{i=1}^{K} \sum_{\gamma \in p(\alpha)} J_i^\gamma \sigma_i + N H(m(\sigma, \mathbf{B}))}$$
$$(39)$$

with ν_α defined as in.[26] Introducing again the integral representation of $\delta_{m(\sigma),m}$ we get

$$\tilde{\alpha}(\beta, t = 0) = \lim_{N \to \infty} \frac{1}{N} \mathbb{E} \log \int_{-1}^{1} dm \int_{-\infty}^{\infty} \frac{d\hat{m}}{\sqrt{2\pi N}} e^{N H(m, \mathbf{B}) + N i m \hat{m} + N f(\hat{m})}$$
$$(40)$$

with $f(\hat{m}) = \mathbb{E} \log \sum_{\alpha} \nu_\alpha e^{\log \cosh(\beta \sum_\gamma J^\gamma - i\hat{m})} = \log 2 + f(0, -i\hat{m}; x(\mathbf{Q}), \beta)$. Using the saddle point tecnique we finally get

$$\tilde{\alpha}(\beta, t = 0) = \text{Extr}_{m,\hat{m}} \left[\log 2 + H(m, \mathbf{B}) + i m \hat{m} + f(0, -i\hat{m}; x(\mathbf{Q}), \beta) \right].$$
$$(41)$$

On the extremum point we get, deriving w.r.t m, $-i\hat{m} = \partial_m H(m, \mathbf{B})$ and, deriving w.r.t. $\hat{m}$, $-im = \mathbb{E}[f_1 \dots f_K \partial_{\hat{m}} \log \cosh(\beta \sum_{l=1}^{K} J_l - i\hat{m})]$. $\square$

Remark 2. We stress that without the introduction of the first-order (ferromagnetic and anti-ferromagnetic) fluctuation-source $H(m, \mathbf{B})$, $\alpha(\beta, t = 0)$ would be $\mathbf{B}$-independent, as the interpolating structure loses the correlation among fields $\{J^l\}_{l=1}^K$. As a consequence, deriving w.r.t. B_l affects $H(m, \mathbf{B})$ only.

We can finally evaluate the self-consistencies which must be respected by the optimal order parameters $(\bar{\mathbf{Q}}, \bar{\mathbf{B}})$ in the thermodynamic limit. We have

$$\partial_{B_l} \alpha(\beta, t = 0) = \partial_{B_l} H(\bar{m}, \mathbf{B}) - \bar{m} \partial_{B_l} \partial_{\bar{m}} H(\bar{m}, \mathbf{B}) + \bar{m} \partial_{B_l} \partial_{\bar{m}} H(\bar{m}, \mathbf{B}),$$

hence

$$\partial_{\bar{B}_l} H(\bar{m}, \mathbf{B})|_{\bar{\mathbf{B}}} = 0. \qquad (42)$$

A. Galluzzi, D. Tantari, F. Guerra

Now, despite physically intuitive, to conclude the proof of the generalized correlated-universality even in the broken-replica-phase, it is left to rigorously prove that the system never displays a not-zero correlation. We tackle this problem, at first, from the simplest perspective of a source term which is positive by definition (it is a square), then we enlarge the proof to the general case.

If we analyze the test-case $B_K^2 = \sum_{l=1}^{K}(m_{l+1}-m_l)B_l^2$, or in other words $B_l = B_K$, $\forall l \in (1,...,K) \Rightarrow b_1 = b$, $b_{l>1} = 0$, then

$$H(\bar{m}, b) = \frac{\beta^2 c}{4}\left(\bar{m}^2 - \frac{b}{c}\right)^2 \Rightarrow \partial_b H(\bar{m}, b) = 0 \Rightarrow \bar{m}^2 = \frac{b}{c}, \quad (43)$$

that perfectly mirrors the replica-symmetric universality as it implies $\partial_{\bar{m}} H(\bar{m}, b) = 0$, hence $\bar{m} = \mathbb{E}\tanh(\beta \sum_{l=0}^{K} J_l) = 0$, thus $\bar{b} = \bar{m} = 0$ and both the fluctuation's source and the quenched free energy approach the standard fully broken Sherrington-Kirkpatrick ones in the thermodynamic limit.[9,11,21]

The general case can be obtained by noticing that as

$$H(\bar{m}, \mathbf{B}) = \frac{\beta^2 c}{4}\left(\bar{m}^4 - 2\frac{B^K}{c}\bar{m}^2 + \sum_{l=1}^{K}(m_{l+1}-m_l)\left(\frac{B_l}{c}\right)^2\right), \quad (44)$$

$$\partial_{B_s} H(\bar{m}, \mathbf{B}) = \frac{\beta^2 c}{4}\left(-\delta_{s,K}\frac{2}{c}\bar{m}^2 + 2(m_{s+1}-m_s)\left(\frac{B_s}{c^2}\right)\right) = 0, \quad (45)$$

hence $\bar{B}_s = 0$, $\forall s < K$ and $\bar{B}_K = \frac{\bar{m}^2 c}{m_{K+1}-m_K}$ that, once plugged into the self-consistency implies

$$H(\bar{m}) = \frac{\beta^2 c}{4}\left(\bar{m}^4 - \frac{2\bar{m}^4}{(m_{K+1}-m_K)} + \frac{\bar{m}^4}{(m_{K+1}-m_K)}\right)$$

$$= \frac{\beta^2 c}{4}\bar{m}^4\left(1 - \frac{1}{(m_{K+1}-m_K)}\right) < 0$$

by which we can conclude that

$$\bar{m} = \left\langle \tanh(\beta \sum_{l=1}^{K} J_l + \partial_{\bar{m}} H(\bar{m}))\right\rangle_K = 0 \quad (46)$$

and we re-obtain the fully-broken free-energy of the standard Sherrington-Kirkpatrick model, proving Theorem 1.

5. Conclusions and outlooks

In this paper we analyzed a generalized universality in a mean field spin glass model for weak coupling correlations $\mathbb{E}(J_{ij}J_{kl}) = c/N^2$, in the case of Gaussian quenched noise. This result generalizes the standard universality of Carmona,[15] proven in the framework of uncorrelated quenched couplings. We have shown that the quenched free energy, and thus the equilibrium states, of a mean field spinglass system, with gaussian symmetric distributed but weakly correlated couplings, does depend only on the variance $\mathbb{E}(J_{ij}^2)$. In the case of positive correlation the result follow easily from the representation of the couplings in term of a common vanishing ordered term, as shown in Section 2. In the case of negative correlations, such a representation no longer holds but it was still possible to derive the free energy of the model, in the RS (Section 3) and RSB (Section 4) approximation, through standard interpolating techniques, in terms of the well-known Parisi variational principle, as in the original SK model.

Acknowledgments

The authors are grateful to Michele Castellana and Dmitry Panchenko for highlighting discussions, GNFM (Gruppo Nazionale per la Fisica Matematica) and INFN (Istituto Nazionale Fisica Nucleare). A.G. and D.T. are partially supported by "Avvio alla Ricerca 2014", Sapienza University of Rome that is acknowledged too.

References

1. E. Agliari, A. Annibale, A. Barra, A.C.C. Coolen, D. Tantari, *J. Phys. A* **46**(33), 335101 (2013).
2. E. Agliari, A. Annibale, A. Barra, A.C.C. Coolen, D. Tantari, *J. Phys. A* **46**(41), 415003 (2013).
3. E. Agliari, A. Barra, G. Del Ferraro, F. Guerra, D. Tantari, *Anergy in self-directed B lymphocytes: A statistical mechanics perspective*, in press in J. Theor. Biol. *doi.org*/10.1016/*j.jtbi*.2014.05.006, (2014).
4. E. Agliari, A. Barra, A. Galluzzi, F. Guerra, D. Tantari, F. Tavani, *J. Phys. A: Math. Theor.* **48** 015001 (2015).
5. E. Agliari, A. Barra, A. Galluzzi, F. Guerra, D. Tantari, F. Tavani, *Phys. Rev. Lett.* **114**, 028103 (2015).
6. G. Ben Arous, A. Dembo, A. Guionnet, *Prob. Theor. and Relat. Fiel.* **120**(1), 1 (2001).
7. A. Barra, G. Genovese, F. Guerra, D. Tantari, *J. Stat. Mech.* **P07**, 009 (2012).

8. A. Barra, P. Contucci, E. Mingione, D. Tantari, *Annals H. Poincaré* **16**(3), 691 (2015).
9. A. Barra, A. Di Biasio, F. Guerra, *JSTAT* **P09**, 006 (2010).
10. A. Barra, G. Genovese, F. Guerra, D. Tantari, *J. Phys. A* **47**, 155002 (2015).
11. A. Barra, F. Guerra, E. Mingione, *Phil. Mag.* **92**(1-3), 78 (2012).
12. A. Barra, G. Genovese, F. Guerra, *J. Stat. Phys.* **140**, 784 (2010).
13. A. Bovier, *Statistical mechanics of disordered system. A mathematical perspective.* Cabridge University Press, Cambridge (2006).
14. M. Castellana, A. Barra, F. Guerra, *J. Stat. Phys.* **155**(2), 211 (2014).
15. P. Carmona, Y. Hy, *Ann. Inst. H. Poincare' (B)* **42**, 215 (2006).
16. W.K. Chen, *Ann. Inst. H. Poinc.* **50**, 1 (2014).
17. P. Contucci, M. Degli Esposti, C. Giardina', S. Graffi, *Comm. Math. Phys.* **236**(1), 55, (2003).
18. A.C.C. Coolen, R. Kuehn, P. Sollich, *Theory of neural information processing systems.* Oxford University Press (2005).
19. H.K. Fischer, J. A. Hertz, *Spin glasses.* Cambridge University Press (1993).
20. F. Guerra, *Int. J. Mod. Phys. B* **10**, 1675 (1996).
21. F. Guerra, *Comm. Math. Phys.* **233**, 1 (2003).
22. G. Genovese, *J. Math. Phys.* **53**, 123304 (2012).
23. G. Genovese, D. Tantari, *Legendre Duality of Spherical and Gaussian Spin Glasses*, arXiv:1409.0062 (2015)
24. J.J. Hopfield, *Proc. Nat. Acad. Sci. USA* **79**, 2554 (1982).
25. M. Mézard, G. Parisi and M.A. Virasoro, *Spin glass theoryand beyond.* World Scientific, Singapore (1987).
26. D. Panchenko, *The Sherrington-Kirkpatrick model.* Springer, 2013.
27. D. Panchenko, *The free energy in a multi-species Sherrington-Kirkpatrick model*, arXiv:1310.6679, to appear in Ann. Probab.
28. P. Sollich, D. Tantari, A. Annibale, A. Barra, *Phys. Rev. Lett.* **113**, 238106 (2014).
29. M. Talagrand, *Spin glasses: A challenge for mathematicians. Cavity and mean field models.* Springer-Verlag (2003).
30. M. Talagrand, *Ann. of Math.* **221**, 263 (2006).
31. L. Viana, A.J. Bray, *J. Phys. C: Solid State Physics*, **18**(15), 3037 (1985).
32. P. Young, *Comp. Sim. in Cond. Matt.* **2**, 31 (2006).

Chapter 5

Poker cash game: A thermodynamic description

M.A. Javarone

Mathematics and Computer Science Department, University of Cagliari (Cagliari, Italy); Humanities and Social Science Department, University of Sassari (Sassari, Italy)

Poker is one of the most popular card games, whose rational investigation represents also one of the major challenges in several scientific areas, spanning from information theory and artificial intelligence to game theory and statistical physics. In principle, several variants of Poker can be identified, although all of them make use of money to make the challenge meaningful and, moreover, can be played in two different formats: tournament and cash game. An important issue when dealing with Poker is its classification, i.e., as a 'skill game' or as gambling. Nowadays, its classification still represents an open question, having a long list of implications (e.g., legal and healthcare) that vary from country to country. In this study, we analyze Poker challenges, considering the cash game format, in terms of thermodynamics systems. Notably, we propose a framework to represent a cash game Poker challenge that, although based on a simplified scenario, allows both to obtain useful information for rounders (i.e., Poker players), and to evaluate the role of Poker room in this context. Finally, starting from a model based on thermodynamics, we show the evolution of a Poker challenge, making a direct connection with the probability theory underlying its dynamics and finding that, even if we consider these games as 'skill games', to take a real profit from Poker is really hard.

1. Introduction

Today understanding Poker games, from the mental aptitide of their players to the underlying probabilistic structure, represents a great challenge for scientists belonging to several communities as psychologists, computer scientists, physicists, and mathematicians.[1-4] In general, these games can be analyzed considering psychological aspects, information theory approaches and analytical descriptions. Notably, approaches based on sociophysics[5–10] allow to study the role of human behavior.[3,11] On the other hand, information theory and analytical approaches allow to identify both new algorithms[4,12] in the context of artificial intelligence,[13] and universal properties of these games.[2] An interesting problem, when dealing with Poker, is constituted by its classification, i.e., 'skill game' or gambling. This issue has not yet been solved, although the nowadays available related answer has a long list of implications.[14,15] A preliminary attempt to solve this question, by using the framework of statistical mechanics, has been developed in,[3] where the author analyzed the role of rationality in a simplified scenario, referred to the Poker variant called Texas Hold'em.[16] In general, all variants follow a similar logic: rounders (i.e., Poker players) receive a number of cards, and have to decide if to bet or not, by computing the possible combinations they can set with their cards (called *hand*). After evaluating if the received hand is promising or not, each rounder can take part to the pot by placing a bet (money or chips), otherwise she/he folds the *hand*. Therefore, the use of money makes the challenge meaningful, otherwise none would have a reason to fold her/his hand. Poker challenges can follow two different formats, i.e., cash game or tournament. During a tournament, rounders pay, only once, an entry fee: a fraction goes into the prize pool, and the remain part is a fee to play. Eventually, top players share the prize pool (usually money). On the other hand, playing Poker in the cash game format means to use real money during the challenge. In this case, rounders can play until they have money and, although there no entry fees to pay, a fraction of each pot is taxed, i.e., a small 'rake' is applied. In this work, we propose a framework to study the evolution of a Poker challenge, considering the cash game format, by a thermodynamic description. In particular, we aim both to model these dynamics to achieve insights, and to link the resulting thermodynamic description with the probability theory tacitly governing these games.

2. Mapping Poker to a Thermodynamic System

In this work, we aim to describe cash game Poker challenges by the language of thermodynamics. In particular, since these challenges entail transfers of money among different parts, i.e., rounders and dealers, we assume that the way thermodynamics explains equilibria and energy transfers between systems constitutes a fundamental tool to our investigations. Firstly, we consider a simple thermodynamic system composed of the subsystem S

and its environment R. The total energy of the system E_T is given by the energy of S and that of R, i.e., $E_T = E_S + E_R$. In the proposed model, the environment R is a Poker room, whereas S corresponds to the table where two rounders, say A and B, face by a 'heads-up' challenge. A 'heads-up' is a challenge characterized by the presence of only two rounders. Therefore, we can identify two subsystems of S: S_A and S_B, corresponding to the two rounders. Since Poker challenges are performed following the cash game format, the money is the exchanged quantity, hence mapping money to the energy of the systems come immediate. In doing so, we have E_A and E_B that correspond to the money of A and B, respectively. Therefore $E_S = E_A + E_B$ and, as initial condition, we impose that at $t = 0$ rounders have the same amount of money, i.e., $E_A(0) = E_B(0)$. Figure 1 offers a pictorial representation of the described system. During the challenge,

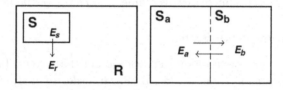

Fig. 1. Thermodynamic representation of a cash game Poker challenge. On the left, the subsystem S inside the environment R, with the arrow indicating the allowed direction of energy transfers, i.e., from S to R. On the right, a zoom on the subsystem S, showing the two subsystems S_A and S_B, representing the two rounders (i.e., A and B). Inside S, as shown by arrows, the energy can flow from S_A to S_B and vice versa.

some rounds are won by A and others by B; hence a fraction of energy is transferred from the subsystem S_A to S_B, and vice versa, over time. The amount of transferred energy Δ_S corresponds to the total amount of money that flows from A to B and vice versa. For the sake of simplicity, we consider that at each round rounders bet the same amount of money, i.e., pots are constant. In particular, Δ_S is defined as

$$\Delta_S(t) = \Phi_{A,B}(t) + \Phi_{B,A}(t), \qquad (1)$$

with $\Phi_{x,y}(t)$ total flow of energy from the subsystem x to y, at time t. Then, $\Phi_{A,B}$ indicates the total amount of enegy transferred from S_A to S_B, as result of all successes of the rounder B. It is worth to note that, in real scenarios, Poker rooms apply a small fee, called 'rake', to each pot. Usually, the 'rake' corresponds to about 5% of the pot. As result, the total energy of S decreases over time by a factor $\Delta_S \cdot \epsilon$, due to energy transfers between S_A and S_B. Since we are dealing with a closed system (i.e., S), the loss in energy can be thought in terms of energy reduction due to the entropy's growth σ. Notably, this concept characterizes the Helmholtz free energy potential F

$$F = E - T\sigma, \tag{2}$$

with E and T, energy and temperature of the system, respectively. In few words, the free energy corresponds to the energy a system can actually use. Then, we can map this concept to our model by the following relation

$$F_S(t) = E_S(t) - \Delta_S(t) \cdot \epsilon, \tag{3}$$

with $F_S(t)$ free energy of our system, that is available at time t, after all energy transfers. The energy lost by S goes to the environment R, hence $E_R = \Delta_S(t) \cdot \epsilon$. It is worth to observe that, as the entropy of a system does, the quantity $\Delta_S(t) \cdot \epsilon$ increases over time, and can never be negative.

2.1. *Evolution of the System*

Now, we focus our attention on the evolution of the system. In particular, we consider one subsystem, i.e., S_A or S_B, in order to analyze its amount of energy over time. Let us consider, for instance, S_A (that represents the rounder A) whose evolution can be described by the following relation

$$E_A(t) = E_A(0) - \Phi_{A,B}(t) + \Phi_{B,A}(t) \cdot (1 - \epsilon), \tag{4}$$

as the amount of energy in S_A corresponds to the initial amount of energy in this subsystem (i.e., $E_a(0)$), minus the amount of energy that flowed to S_B (i.e., $\Phi_{A,B}(t)$), plus the amount of energy that flowed from S_B to S_A (i.e., $\Phi_{B,A}(t)$) reduced of a factor ϵ. In an equilibrium condition, $\lim_{t \to \infty} E_A(t) = E_A(0)$, therefore

$$\Phi_{A,B}(t) = \Phi_{B,A}(t) \cdot (1 - \epsilon). \tag{5}$$

The primary target of the rounder A is to win all the money of B, while avoiding to lose her/his money. Hence, rounder A aims to obtain $E_A(t) \geq E_A(0)$. Considering Poker as a skill game,[3] the flow $\Phi_{x,y}$ depends on the ability of the yth rounder. Then, A has a probability P_a to win each round, strongly related to her/his skills. As discussed before, to simplify the scenario, we suppose that rounders bet always the same amount of money $\frac{\delta}{2}$, forming the pot δ, so that $\Delta_S(t) = \delta \cdot t$. In doing so, we can define the amount of energy transferred from S_B to S_A as

$$\Phi_{B,A} = P_a \cdot \Delta_S \tag{6}$$

and the amount of energy transferred from S_A to S_B

$$\Phi_{A,B} = (1 - P_a) \cdot \Delta_S. \tag{7}$$

Going back to the equilibrium condition defined in Equation (5), we can write

$$P_a \cdot \Delta_S \cdot (1 - \epsilon) = (1 - P_a) \cdot \Delta_S, \tag{8}$$

working a little bit of algebra, from Equation (8), we obtain

$$P_a = \frac{1}{2 - \epsilon}. \tag{9}$$

Therefore, we can compute the minimal success probability that the rounder A needs to reach her/his target, i.e., to win (or, at least, to not losing money). It is worth to highlight that, starting by a thermodynamic description of the system, we can define a relation between the 'rake', applied by a Poker room, and the rounders' skills, i.e., their probability to success in Poker cash game.

2.2. *Profits over time*

In light of these results, it is interesting to evaluate both the amount of money rounders can win by playing Poker cash game, and the amount of money the Poker room generates during the challenge. Considering the rounders's perspective, we are interested in computing the expected value of energy that flows in the subsystem S_A, i.e., $< \Delta E_A >$ (note that similar considerations hold also for S_B). The value of $< \Delta E_A >$ can be computed as follows

$$< \Delta E_A > = < \Phi_{B,A} > \cdot (1 - \epsilon) - < \Phi_{A,B} >, \tag{10}$$

$< \Phi_{B,A} >$ corresponds to $< \Phi_{B,A} > = P_a \cdot \varphi_{B,A}$, and $< \Phi_{A,B} > = (1 - P_a) \cdot \varphi_{A,B}$, with $\varphi_{x,y}$ representing the total energy transfer from S_x to S_y, i.e., $\varphi = t \cdot \frac{\delta}{2}$. Then, we obtain

$$< \Delta E_A > = P_A \cdot t \cdot \frac{\delta}{2} \cdot (1 - \epsilon) - (1 - P_A) \cdot t \cdot \frac{\delta}{2} \tag{11}$$

and we find the following relation

$$< \Delta E_A > = t \cdot \frac{\delta}{2} [P_A \cdot (1 - \epsilon) - 1 + P_A] = t \cdot \frac{\delta}{2} [P_A \cdot (2 - \epsilon) - 1], \tag{12}$$

that is in perfect accordance with results achieved in Equation (9), as for $P_a = \frac{1}{2-\epsilon}$ the expected value of energy transferred to S_A is $< \Delta E_A > = 0$.

Moreover, it is immediate to note that, considering Equation (4), the rate
of variation of the energy of one subsystem (e.g., S_x) corresponds to

$$\frac{dE_x}{dt} = \frac{<\Delta E_x>}{t}.$$ (13)

It is worth to observe that, by all the illustrated equations, it is possible to
evaluate the potential gain of a rounder, once her/his winning probability
P_x is known. On the other hand, considering the Poker room perspective,
the overall scenario becomes pretty nice, because as we are going to show,
its profits can only increases over time without running any risk. Notably,
the environment R (i.e., the Poker room) receives a constant amount of
energy, at each time step, equal to $\delta \cdot \epsilon$. Hence, in the event rounders have
the same probability to win (i.e., $P_A = P_B$), it is interesting to compute
the number of time steps required to let the Poker room drain almost all
their money (i.e., their energy). Since to perform a round both rounders
have to bet the same amount of money, a minimal amount of energy always
will remain in the subsystem S. In particular, this quantity is equal to $\frac{\delta}{2}$.
Hence, supposing both subsystems, at time t, contain an energy equal to
$\frac{\delta}{2}$, the last round entails one subsystem loses completely energy and the
other has, at the end, an energy equal to $\delta(1-\epsilon)$. Therefore, the maximum
amount of energy that the environment can receive is $2 \cdot E_s(0) - \delta(1-\epsilon)$,
so that the following relation holds

$$\Delta_s \epsilon t = 2E_s(0) - \delta(1-\epsilon),$$ (14)

then, it is possible to compute the number of time steps t to let the Poker
room draining almost all the rounders's money:

$$t = \frac{2E_s(0) - \delta(1-\epsilon)}{\Delta_s \epsilon}.$$ (15)

Equation (15) shows a direct relation between the time and the rake applied
by the Poker room: as the latter increases the time to drain almost all the
energy decreases.

3. Discussion and Conclusions

In this work, we propose a framework for studying Poker challenges in the
context of thermodynamics. In particular, we map a simple scenario, where
two rounders face, to a thermodynamic system composed of a subsystem
S embedded in a larger environment R. The former represents the two
rounders, whereas the latter the Poker room. Remarkably, from a simplified

description of the game dynamics, we achieve insights on Poker challenges, in the cash game format. Even considering this format of Poker as a 'skill game' (see[3]), we identify a direct link between the rounders's skills and the fee applied by Poker room, called 'rake'. In doing so, it is possible to know the minimal probability to success a rounder needs to have in order to be a successful player. As shown, gaining by this activity is a very hard task, even for skilled rounders, as they have to keep their probability to win very high. In real scenarios, many expert rounders are very good and fast in computing winning probabilities for their *hands*, hence they perform online cash game by a 'multitabling' strategy: they face at the same time several opponents, with the aim to optimize their profits (obviously, increasing the probability of losing a lot of money). Moreover, we analyze profits of a Poker room, obtained while rounders play the cash game Poker. In particular, mapping this profit to the energy of the environment R, we achieve the relation $E_R(\epsilon, t) = \delta \epsilon t$, with ϵ representing the 'rake', δ the pot of each round, and t the number of time steps. It is worth noting that there are two different situations that allow the Poker room to increase its profits:

(1) Increasing ϵ
(2) Increasing t

While the first should be kept low (e.g., 5% or less) as a strategy marketing to attract rounders in the Poker room, the second requires more attention as, in principle, it can lead to a fraudulent strategy, now we briefly illustrate. People usually are not worried about frauds in Poker, as they play against other people, and not against a dealer (as in games like the roulette). Therefore, in principle, there are no reasons for the electronic dealer to favor a particular rounder in the process of cards distribution. Anyway, it is important to highlight that the Poker room does not take an advantage when rounders perform 'all-in' actions (i.e., the bet all their money in only one *hand*). Then, supposing rounders are rational, i.e., their actions are performed by considering their probability to win each round, a pseudo-random algorithm for cards distribution could be properly defined for generating uncertain scenarios. Here, for uncertain scenarios we indicate those situations where both rounders have low winning probabilities, by considering only the information they have (i.e., their *hand* and, in case, common cards). Therefore, a fraudulent strategy could be implemented by using an algorithm to provide rounders with low winning probability at each hand, in order to avoid they perform 'all-in'. It is evident that by this strategy, it would be possible to indirectly increasing t for each challenge. Moreover, it would be also very difficult to find this kind of fraud by analyzing the algorithm, used by a Poker room, if this fraudulent scenario is not considered. In order to conclude, we would like to emphasize that some of the considerations about the probability to win a cash game challenge can be applied also in the context of financial trading. In particular, for the strategy adopted by 'scalpers', i.e., traders that in few seconds open and close a position (i.e., buy and sell financial products as stocks, bonds, etc.), as also in those cases for each transaction the banking system applies a kind of 'rake'.

References

1. M. Bowling, N. Burch, M. Johanson, O. Tammelin, *Science* **347** 145–149 (2015).
2. C. Sire, *JSTAT* P08013 (2007).
3. M.A. Javarone, *JSTAT* in press (2015).
4. L.F. Teofilo, L.P. Reis, H. Lopes Cardoso, *Inform. Sys. Tech. (CISTI), 2013 8th Iberian Conference on* 1–6 (2013).
5. A. Barra, P. Contucci, R. Sandell, C. Vernia, *Nature Sci. Rep.* **4**, 4174 (2014).
6. E. Agliari, A. Barra, P. Contucci, R. Sandell, C. Vernia, *New J. Phys.* **16**, 103034 (2014).
7. E. Agliari, A. Barra, A. Galluzzi, M.A. Javarone, A. Pizzoferrato, *to appear in PLoS-One, available at arXiv:1503.00659* (2015).
8. S. Galam, *International Journal of Modern Physics C* **19**, 409 (2008).
9. C. Castellano, S. Fortunato, V. Loreto, *Rev. Mod. Phys.* **81**, 591 (2009).
10. M.A. Javarone, *Physica A* **414**, 19 (2014).
11. M.G. Kim, K. Suzuki *Int. J. Soc. Robot.* **1**, 5 (2013).
12. F.A. Dahl, *Mach. Learn.* **2167**, 85 (2001).
13. S. Russell, P. Norvig, Artificial Intelligence: a Modern Approach (3rd edition) *Pearson* (2009).
14. J.M. Kelly, Z. Dhar, T. Verbiest, *Gaming Law Rev.* **3** (2007).
15. A. Cabot, R. Hannum, *TM Cooley L. Rev.* **22** (2005).
16. D. Sklansky, M. Malmuth, Hold 'em Poker for Advanced Players. *Two Plus Two Publications* (1999).

Chapter 6

Accretion processes in astrophysics: Cross-fertilization with laboratory plasmas?

Giovanni Montani

ENEA - Fusion Technical Units, R.C. Frascati (Frascati (Rome), Italy);
Physics Department, "Sapienza" University of Rome (Rome, Italy)

Riccardo Benini

Physics Department, "Sapienza" University of Rome (Rome, Italy)

Nakia Carlevaro

ENEA - Fusion Technical Units, R.C. Frascati (Frascati (Rome), Italy)

We analyze in some details the morphology of the equilibrium configuration characterizing a plasma disk surrounding an astrophysical compact object and subjected to the axial symmetry restriction. The work reviews a number of relevant features concerning the different implications of the momentum conservation component, mainly in the presence of dissipation effects.

We start by re-analyzing the Shakura standard model for thin accretion disks, clarifying the role of the effective shear viscosity in allowing the angular momentum transport toward the central object. Then, we discuss a two-dimensional reformulation of the problem, which relies on the emergence of a small scale backreaction in the plasma. We set up the basic equilibrium configuration equations and provide a solution of the linear regime, when the induced magnetic field remains smaller than the background one. This approach, based on the idea of treating the plasma disk as a quasi-ideal medium, applies in the limit of small poloidal velocities and it is derived on a local level, nearby a fiducial radius of the thin configuration.

We generalize such a picture to a full global profile, for which the radial dependence of the background enters the perturbation dynamics, but in the absence of poloidal velocities of the plasma. The obtained scenario clarifies how, in the limit of an extreme non-linear backreaction of the plasma, the radial oscillation of the magnetic flux function (already outlined in the local linear model) becomes a dominant effect in fixing the magnetic structure of the disk and hence produces a decomposition of the disk into a ring series, the really new feature of the two-dimensional revised scheme.

Finally, we outline the cross fertilization character, between laboratory and astrophysical plasma, of such new paradigm, elucidating a qualitative proposal to explain the accretion properties of a quasi-ideal plasma disk in terms of the plasma porosity, nearby the X-point of the magnetic configuration, due to the crystalline profile induced by the backreaction.

1. Introduction

In the limit of a thin gaseous disk, the hydrodynamical equilibria underlying the accretion process on a compact astrophysical object can be set up as a 1D paradigm (see among the first analyses of the problem[1-3]). In such a fluidodynamical scenario, the accretion mechanism relies on the angular momentum transfer as allowed by the shear viscosity properties of the disk material. The differential angular rotation of the radial layers is associated with a non-zero viscosity coefficient, accounting for diffusion and turbulence phenomena. Indeed, microscopic estimations of the viscosity parameter indicate that the friction of different disk layers would be unable to maintain a sufficiently high accretion rate, and therefore non-linear turbulent features of the equilibrium are inferred. However, when the magnetic field of the central object is strong enough, the electromagnetic back-reaction of the disk plasma becomes relevant.[4] As shown in,[5] the Lorentz force induces a coupling between the radial and the vertical equilibria, which deeply alters the local morphology of the system; the radial dependence of the disk profile acquires, in particular, an oscillating character, modulating the background structure too. The existence of such a coupling breaks down the 1D nature of the problem and suggests a revision of the original standard model.[6]

This revised paradigm, based on an ideal magneto-hydrodynamics (MHD) approach, constitutes an opposite point of view with respect to the idea of a diffusive magnetic field within the disk, as discussed in,[7] since it relies on the existence of magnetic micro-structures in radial disk morphology. Here, we present a detailed comparison between the 1D standard model and the 2D ideal MHD reformulation, which is analyzed both in the local[5,8,9] and global[10] formulation of the problem. We treat the backreaction process in the linear limit when the induced magnetic field inside the disk remains small compared with the background one. Only in the last part of the present paper we trace the basic ideas for a revised ideal accretion process, stressing how the emergence of a ring-series decomposition of the disk is a crucial feature. We provide a discussion of the difficulties associated with including velocity fields in the steady 2D equilibrium, and clarify how a net non-zero accretion rate is not directly inferred from this revised approach; we finally stress how a cross-fertilization between the laboratory and astrophysical plasma could be the only proper way to get a real accretion feature in an ideal plasma configuration.

The paper is structured as follows. In Sec. 2, a discussion of the basic equations in the steady MHD framework is provided. In Sec. 3, the standard 1D Shakura paradigm for the accretion rate *versus* viscosity is summarized. In Sec. 4, the 2D local MHD model for an accretion disk is discussed in some details. A solution in the limit of very low back-reaction is presented in Sec. 5. In Sec. 6, the extension of the MHD framework to a global disk profile is presented, and the perturbation scheme is analyzed in some details demonstrating the appearance of a ring series decomposition. In Sec. 7, a hint for a possible explanation of the accretion mechanism is given within the crystalline structure paradigm. Concluding remarks follows.

2. Stationary MHD theory

The behavior of a steady plasma configuration embedded into a magnetic and a gravitational fields is governed by the following set of equations in Cartesian coordinates (repeated indices are intended as summed):

$$\partial_i \left(\epsilon v_i \right) = 0 \,, \tag{1a}$$

$$\epsilon v_l \partial_l v_i = -\partial_i p + \partial_l [\mathcal{D} \left(\partial_i v_l + \partial_l v_i - 2\delta_{il}\partial_k v_k/3 \right)]$$
$$-\partial_i \chi + \epsilon_{ijk} J_j B_k/c \,, \tag{1b}$$

$$v_l \partial_l T + 2T\partial_l v_l/3 = 0 \,, \tag{1c}$$

where ϵ is the mass density and p the total pressure and they are related via the equation of state $p = 2K_{\mathrm{B}}T\epsilon/m$ (T being the temperature, K_{B} the Boltzmann constant and m the proton mass); $\mathcal{D}$ denotes the shear viscosity coefficient, J_i the density current vector, v_i the plasma velocity field, and χ stands for the Newton potential due to the central body of mass M_{S} (the self-gravity of the plasma being negligible), i.e. $\chi(x^i) = GM_{\mathrm{S}}/\sqrt{\delta_{ij}x^i x^j}$.

The dynamics of the electric field E_i and of the magnetic field B_i is summarized by the Maxwell equations in Gaussian units as

$$\epsilon_{ijk}\partial_j B_k = 4\pi J_i/c \,, \tag{2a}$$

$$\epsilon_{ijk}\partial_j E_k = 0 \quad \Rightarrow \quad E_i = -\partial_i\Phi \,, \tag{2b}$$

$$\partial_l B_l = 0 \quad \Rightarrow \quad B_i = \epsilon_{ijk}\partial_j A_k \,; \tag{2c}$$

here A_i is the potential vector, Φ the electric potential and the following MHD relation holds (electron force balance)

$$E_l = -\epsilon_{lmn}v_m B_n/c \,. \tag{3}$$

This system corresponds to a closed partial differential problem which fixes the plasma steady configuration surrounding a given compact astrophysical object.

3. Standard accretion model

The description of an accretion structure comes out by addressing a steady axisymmetric MHD configuration (in cylindrical coordinates $[r, \phi, z]$) for the plasma. Nonetheless, some basic features of the accretion mechanism can be fixed by a fluid-dynamical approach describing the matter infall across a gas disk profile.[6] In the standard model of accretion, the configuration of an axisymmetric thin disk is governed by the hydrostatic equilibrium within the central object gravity and the problem is reduced to a 1D paradigm by integration over the vertical disk profile.

Averaging out the vertical dependence,[7] the radial equilibrium essentially reduces to the condition

$$\omega(r) = \omega_K \equiv \sqrt{GM_S/r^3} \,. \tag{4}$$

In other words, we are neglecting the role played in the equilibrium by the pressure radial gradient. The vertical equilibrium fixes the gravothermal profile of the disk, i.e.,

$$dp/dz + \omega_K^2 z\epsilon = 0 \,, \tag{5}$$

which, for an isothermal disk with an equation of state $p = v_s^2\epsilon$ (where $v_s^2 = 2K_B T/m$ is the sound velocity), gives the exponential profile

$$D(z^2) \equiv \epsilon/\epsilon_0 = \exp\left[-z^2/H^2\right] \,, \tag{6}$$

where $\epsilon_0(r)$ is the equatorial mass density and $H \equiv \sqrt{2v_s^2/\omega_K^2}$ estimates the the half-depth of the disk. The azimuthal equilibrium is responsible for the angular momentum transport across the disk, by virtue of a viscous stress tensor component $\tau_{r\phi}$ entering the relation (first integral of the azimuthal equation)

$$\dot{M}_d(L - L_d) = -2\pi r^2 \tau_{r\phi} , \quad \tau_{r\phi} = \mathcal{D} r^2 d\omega/dr . \qquad (7)$$

Here, $\mathcal{D}$ is an effective turbulent viscosity coefficient, L is the specific angular momentum, L_d is the value of L on the disk inner boundary layer and

$$\dot{M}_d \equiv -2\pi r \int_{-H}^{+H} (\epsilon v_r) dz = -2\pi r \sigma v_r \qquad (8)$$

is the mass accretion rate, associated to the radial velocity $v_r < 0$ and to the surface mass density $\sigma \equiv \int_{-H}^{H} \epsilon dz$. Finally, the mass conservation equation

$$\frac{1}{r}\frac{d(r\epsilon v_r)}{dr} + \frac{d(\epsilon v_z)}{dz} = 0 , \qquad (9)$$

once integrated over the vertical direction provides $\dot{M}_d = $ const > 0. Since the model is dominated by the Keplerian feature $\omega \simeq \omega_K$, we get the following expression for the accretion rate

$$\dot{M}_d(L - L_d) = 3\pi \mathcal{D} \omega_K r^2 . \qquad (10)$$

It is worth noting that the shear viscosity here advocated cannot be the kinetic contribution due to the microscopic plasma morphology collisionality. Indeed, as suggested in,[6] we are dealing with an effective viscosity induced by the small scale turbulence arising in the plasma as effect of its instability. Since by definition $L = \omega r^2$, the phenomenological expression for the shear viscosity coefficient is then

$$\mathcal{D} = 2\sigma v_t H/3 , \qquad (11)$$

where v_t is a turbulence velocity, given by $v_t = \alpha v_s$, and α is a free dimensionless parameter taking values less that unity.

4. 2D MHD Model for an accretion disk

The Shakura model we traced above finds appreciable phenomenological confirmations and significant agreement with higher dimensional simulations, but its idea of angular momentum transport requires effective dissipation processes to be justified. While the effective viscosity can be reliable explained by the turbulence associated to the magneto-rotational instability (see[11,12] for a detailed discussion), the emergence of an anomalous resistivity[7] is rather puzzling and call attention for a deeper understanding. More specifically, to get an effective viscosity in the plasma disk, it is necessary to account for a weak magnetic field, typically present sufficiently far from the central object. Nonetheless, as soon as such a field is involved, the corresponding MHD scenario (in particular, the generalized Ohm equation) links its strength to the radial infalling velocity. To account for the accretion rate observed in X-Ray binary systems, a very significant resistive nature of the plasma is required and the mechanism responsible for this phenomenon can not be directly recognized in the turbulence regime.

For this reason, in what follows we analyze an ideal 2D model for the accretion process in order to evaluate if a cross-fertilization between the laboratory and astrophysical plasma can help to account for the angular momentum transport even in the absence of effective dissipation. The present reformulation is still an incomplete paradigm which, however, establish a new point of view for the accretion process outlining the main difficulties to be overcome towards a satisfactory picture.

4.1. *Basic morphology of the 2D equilibrium*

Let us now specify the steady MHD theory to the case of an accretion disk configuration around a compact and strongly magnetized astrophysical object, whose gravitational potential has the form

$$\chi(r,z) = GM_S/\sqrt{r^2 + z^2} \,. \tag{12}$$

Since the axial symmetry prevents any dependence on the azimuthal angle ϕ, the continuity equation takes the explicit form (9), which provides the following matter flux

$$\epsilon\vec{v} = -\frac{1}{r}\partial_z\Theta\vec{e}_r + \epsilon\omega(r,z^2)r\vec{e}_\phi + \frac{1}{r}\partial_r\Theta\vec{e}_z \,, \tag{13}$$

where $\Theta(r,z)$ is an odd function of z in order to get a non-zero accretion rate, i.e.

$$\dot{M}_d = -2\pi r \int_{-H}^{+H} \epsilon v_r dz = 4\pi\Theta(r,H) > 0 \,. \tag{14}$$

The central object magnetic field takes the form

$$\vec{B} = -\frac{1}{r}\partial_z\psi\vec{e}_r + \frac{I}{r}\vec{e}_\phi + \frac{1}{r}\partial_r\psi\vec{e}_z \,, \tag{15}$$

where $\psi = \psi(r,z^2)$ and $I = I(\psi,z)$ denotes the magnetic flux surface function and the axial current, respectively.

We now analyze the plasma equilibrium around a fiducial radial co-ordinates $r = r_0$ when the plasma disk backreactrion is taken into account. Let us now split the mass density and the pressure contributions as $\epsilon = \bar{\epsilon}(r_0, z^2) + \hat{\epsilon}$ and $p = \bar{p}(r_0, z^2) + \hat{p}$, respectively. Analogously, we set $\psi = \psi_0(r_0) + \psi_1(r_0, r - r_0, z^2)$, with $\psi_1 \ll \psi_0$. The quantities $\hat{\epsilon}$, $\hat{p}$ and ψ_1 are due to the currents which rise within the disk and, in general, these corrections are expected to have small amplitude and small spatial scale of variation. Thus, we have the following hierarchy in the space gradient: the first order gradients of the perturbations are of zero-order, while the second order ones dominate.

The corotation theorem[13] prescribes that the angular frequency of the disk must be expressed via the flux function only ($\omega = \omega(\psi)$). From this statement the following decomposition comes out: $\omega = \omega_K + \omega'_0 \psi_1$, where ω_K is the zeroth-order (Keplerian) term and $\omega'_0 = $ const. Such an expression of ω remains valid locally, as far as $(r - r_0)$ is small enough. The profile pf the toriodal current arising in the disk takes the form

$$J_\phi \simeq -c \left(\partial_r^2 \psi_1 + \partial_z^2 \psi_1 \right) / 4\pi r_0 . \tag{16}$$

We moreover stress that the the azimuthal component of the Lorentz force is related to the existence of the function $I(\psi, z)$ and it can be written as

$$F_\phi \simeq \left(\partial_z I \partial_r \psi - \partial_r I \partial_z \psi \right) / 4\pi r_0 . \tag{17}$$

4.2. *Vertical and radial equilibria*

We now fix the equations governing the vertical and the radial equilibrium of the disk separating the background component from the backreaction terms.[5,8] Using the definition $\epsilon_0(r_0) \equiv \epsilon(r_0, 0)$, such a splitting of the MHD equations for the vertical force balance gives

$$D(z^2) \equiv \bar{\epsilon}/\epsilon_0(r_0) = \exp\left[-z^2/H^2 \right] , \qquad H^2 \equiv 4K_B \bar{T}/m\omega_K^2 , \tag{18}$$

$$\partial_z \hat{p} + \omega_K^2 z \hat{\epsilon} - \frac{1}{4\pi r_0^2} \left(\partial_z^2 \psi_1 + \partial_r^2 \psi_1 \right) \partial_z \psi_1 = 0 . \tag{19}$$

Here, the temperature T admits the representation

$$2K_B T \equiv m\frac{p}{\epsilon} = m\frac{\bar{p} + \hat{p}}{\bar{\epsilon} + \hat{\epsilon}} \equiv 2K_B(\bar{T} + \hat{T}) . \tag{20}$$

The radial equation can instead be decomposed as

$$\omega \simeq \omega_K + \delta\omega \simeq \omega_0(\psi_0) + \omega'_0 \psi_1 , \tag{21}$$

$$2\omega_{\mathrm{K}}r_0(\bar{\epsilon}+\hat{\epsilon})\omega_0'\psi_1 + \frac{1}{4\pi r_0^2}\left(\partial_z^2\psi_1 + \partial_r^2\psi_1\right)\partial_r\psi_1$$
$$= \partial_r\left[\hat{p} + \frac{1}{8\pi r_0^2}\left(\partial_r\psi_1\right)^2\right] + \frac{1}{4\pi r_0^2}\partial_r\psi_1\partial_z^2\psi_1 , \tag{22}$$

where we have neglected the presence of the poloidal currents, associated with the azimuthal component of the magnetic field.

Let us define the dimensionless functions Y, $\hat{D}$ and $\hat{P}$:

$$Y \equiv \frac{k_0\psi_1}{\partial_{r_0}\psi_0} \qquad \hat{D} \equiv \frac{\beta\hat{\epsilon}}{\epsilon_0} , \qquad \hat{P} \equiv \beta\frac{\hat{p}}{p_0} , \tag{23}$$

where $p_0 \equiv 2K_{\mathrm{B}}\bar{T}\epsilon_0$ and $\beta \equiv 8\pi p_0/B_{0z}^2 = 1/(3\varepsilon_z^2) \equiv k_0^2 H^2/3$. Here we introduced the fundamental wavenumber k_0 of the radial equilibrium, corresponding to $k_0 \equiv 3\omega_{\mathrm{K}}^2/v_{\mathrm{A}}^2$, with $v_{\mathrm{A}}^2 \equiv B_{z0}^2/4\pi\epsilon_0$ being the Alfvén velocity, and recalling that $B_{z0} = B_z(r, z = 0) = \partial_{r_0}\psi_0/r_0$. Furthermore, we introduce the dimensionless radial variable $x \equiv k_0(r - r_0)$, while we assume that the fundamental length in the vertical direction is $\Delta \equiv \sqrt{\varepsilon_z}H$, leading to define $u \equiv z/\Delta$. Using these definitions, the vertical equilibrium rewrites

$$\partial_{u^2}\hat{P} + \epsilon_z\hat{D} - 2\left(\partial_x^2 Y + \epsilon_z\partial_u^2 Y\right)\partial_{u^2}Y = 0 , \tag{24}$$

while the radial equilibrium stands as

$$(D + \hat{D}/\beta)Y + (\partial_x^2 Y + \epsilon_z\partial_u^2 Y)(1 + \partial_x Y) + \partial_x\hat{P}/2 = 0 . \tag{25}$$

The two equations above provide a coupled system for $\hat{P}$ and Y once the quantities D and $\hat{D}$ are assigned; thus we are able to determine the disk configuration due to the toroidal currents. We emphasize that, according to the idea that the plasma is quasi-ideal, we regard the poloidal velocities of the plasma as small contribution to the problem. For this reason, we assumed that their impact to the present radial and vertical equilibria is essentially negligible, while their morphology will be unambiguously fixed by the azimuthal equilibrium and the generalized Ohm law.

4.3. The azimuthal equation

The toroidal equilibrium of the disk is summarized by the following expression

$$\epsilon v_r\partial_r(\omega r) + \epsilon v_z\partial_z(\omega r) + \epsilon\omega v_r = \frac{1}{r^2}\partial_r\left(\mathcal{D}r^3\partial_r\omega\right) + \partial_z\left[\mathcal{D}\partial_z(\omega r)\right] + F_\phi . \tag{26}$$

The corotation theorem permits us to rewrite the equation above as

$$
\begin{aligned}
&\epsilon r v_r \partial_r \psi + \epsilon r v_z \partial_z \psi + 2\epsilon v_r \frac{\omega}{\omega'} \\
&= \frac{1}{r^2 \omega'} \partial_r \left(\mathcal{D} r^3 \omega' \partial_r \psi \right) + \frac{1}{\omega'} \partial_z \left[\mathcal{D} r \omega' \partial_z \psi \right] + \frac{F_\phi}{\omega'} \,,
\end{aligned}
\tag{27}
$$

($\omega' = d\omega/d\psi$). Using (15), the l.h.s. of this equation can be restated obtaining

$$
\begin{aligned}
&\epsilon r^2 \left(v_r B_z - v_z B_r \right) + 2\epsilon v_r \frac{\omega}{\omega'} = \\
&\frac{1}{r^2 \omega'} \partial_r \left(\mathcal{D} r^3 \omega' \partial_r \psi \right) + \frac{1}{\omega'} \partial_z \left[\mathcal{D} r \omega' \partial_z \psi \right] + \frac{F_\phi}{\omega'} \,.
\end{aligned}
\tag{28}
$$

In our local model, by virtue of Eq.(17), the azimuthal equation stands at r_0 as

$$
\begin{aligned}
\epsilon r_0^2 \left(v_r B_z - v_z B_r \right) + 2\epsilon v_r \frac{\omega_K}{\omega_0'} &= r_0 \mathcal{D}_0 \left(\partial_r^2 \psi_1 + \partial_z^2 \psi_1 \right) \\
&+ \frac{1}{4\pi r_0^2 \omega_0'} \left[\partial_z I \left(\partial_{r_0} \psi_0 + \partial_r \psi_1 \right) - \partial_r I \partial_z \psi_1 \right] \,,
\end{aligned}
\tag{29}
$$

where $\mathcal{D}_0 = \mathcal{D}(r_0)$. This equation accounts for the angular momentum transport across the disk, allowed by the presence of viscosity in the disk differential rotation.

4.4. *Generalize Ohm law*

When a non-zero resistivity coefficient η is considered, the generalized Ohm law takes th following form

$$
\vec{E} + \vec{v} \wedge \vec{B}/c = \eta \vec{J} \,.
\tag{30}
$$

When the contribution to the current density due to the toroidal component of the magnetic fiel d is negligible, such vector is along $\vec{e}_\phi$ only and the resistive term enters the azimuthal component of the equation above. Therefore, the balance of the Lorentz force in the meridian plane provides the electric field in the form predicted by the corotation theorem, i.e.

$$
\vec{E} = -\vec{v} \wedge \vec{B}/c = -d\Phi/d\psi \, \vec{\nabla}\psi = -\omega \left(\partial_r \psi \vec{e}_r + \partial_z \psi \vec{e}_z \right)/c \,.
\tag{31}
$$

Since the axial symmetry requires $E_\phi \equiv 0$, the ϕ-component of the equation above in the local formulation around r_0 reads as

$$
v_r B_z - v_z B_r = \eta c^2 \left(\partial_r^2 \psi_1 + \partial_z^2 \psi_1 \right)/4\pi r_0 \,.
\tag{32}
$$

Let us now observe that, substituting Eq.(32) in Eq.(29), we arrive to the expression

$$
\begin{aligned}
2\epsilon v_r \frac{\omega_K}{\omega_0'} =& r_0\left(\mathcal{D}_0 - \frac{c^2\eta\epsilon}{4\pi}\right)(\partial_r^2\psi_1 + \partial_z^2\psi_1) \\
&+ \frac{1}{4\pi r_0^2\omega_0'}\left[\partial_z I\left(\partial_{r_0}\psi_0 + \partial_r\psi_1\right) - \partial_r I \partial_z\psi_1\right].
\end{aligned}
\tag{33}
$$

This relation together with the electron force balance one (32) give a partial differential system for Θ and I, once ψ_1 (i.e. Y) and ϵ (i.e. D and $\hat{D}$) are assigned via the vertical (24) and radial (25) equilibria.

5. The linearized model

We now consider the linear model, corresponding to the condition $\partial_{r_0}\psi_0 \gg \partial_r\psi_1$ and $\partial_z\psi_1 \simeq 0$. These approximations correspond to $Y \ll 1$ in the dimensionless vertical (24) and radial (25) equations, and to the limit $\epsilon_z < 1$ (i.e. $\beta > 1$). In what follows, we also neglect the radial pressure gradient, according to the Keplerian nature of the unperturbed disk.

In this linearized regime, we clearly have $\epsilon = \bar{\epsilon} + \hat{\epsilon} \simeq \bar{\epsilon} = \epsilon_0(r_0)D(z^2)$ and the radial equation reads as

$$
\partial_r^2\psi_1 + \partial_z^2\psi_1 = -k_0^2 D(z^2)\psi_1.
\tag{34}
$$

It is easy to check that the equation above admits the following solution

$$
\psi_1(r, z^2) = A_0 \sin[k(r - r_0)] \exp[-z^2/2\delta],
\tag{35}
$$

where $A_0 \ll 1$ is an integration constant, $k \equiv k_0\sqrt{1 - 1/k_0 H}$ and $\delta \equiv \sqrt{H/k_0}$. Starting from the definition of the β parameter of the plasma, we can recognize the existence of the following relation: $\beta = k_0^2 H^2/3$. Since it is well known that in astrophysical systems β takes very large values for the plasma in accretion (as soon as we are sufficiently far form the central object), we can argue that the magnetic micro-structures emerging in ψ_1 have a small spatial scale, i.e., $k \simeq k_0 \gg 1/H$. This fact provides self-consistency to the backreaction model here proposed, since it justifies the relevant role played by the induced current toroidally flowing in the disk, simply in view of their small spatial scale.

In the same approximation, the linear generalized Ohm law can be easily recast at the lowest order as

$$
v_r \equiv -\frac{1}{r_0\bar{\epsilon}}\partial_z\Theta = \frac{\eta c^2 k_0^2}{4\pi\partial_{r_0}\psi_0}\psi_1,
\tag{36}
$$

and we observe that (as in the dipole magnetic profile) $\partial_{r_0}\psi_0 < 0$.

In order to deal with the linear equation (33), we are naturally led to set the conditions $\eta = \eta_0/D(z^2)$ and $\mathcal{D}_0 = (c^2\eta_0\epsilon_0)/(4\pi)$ (η_0 being the resistivity at r_0), obtaining

$$v_r = \frac{1}{8\pi r_0^2 \omega_K \epsilon_0} \partial_z I \partial_{r_0}\psi_0 . \tag{37}$$

Comparing the two expressions for v_r (36) and (37), we get the compatibility constraint

$$\partial_z I = \frac{2\eta_0\epsilon_0 c^2 (k_0 r_0)^2 \omega_K}{(\partial_{r_0}\psi_0)^2}\psi_1 , \tag{38}$$

providing a relation between the two functions I and ψ_1. Using now (36), we easily get

$$\Theta = -\frac{\eta_0\epsilon_0 c^2 k_0 r_0}{4\pi} \int dz Y . \tag{39}$$

From the above we can determine the vertical and radial velocity components, i.e.

$$v_z \simeq \frac{\partial_r \Theta}{r_0 \epsilon_0} = -\frac{\eta_0 c^2 k_0}{4\pi} \int dz (\partial_r Y) , \qquad v_r \simeq -\frac{\partial_z \Theta}{r_0 \epsilon_0} = \frac{\eta_0 c^2 k_0}{4\pi} Y , \tag{40}$$

where we have neglected the function $D(z^2)$ because its contribution is here of higher order. The main issue of the present analysis of the plasma equilibrium, in the presence of a linear backreaction to the external magnetic field of the central object, is the emergence of a radial oscillation of the magnetic flux function as clearly emerging from Eq.(35).

Before extracting information from this configurational setting towards the accretion properties of a steady plasma disk, in the next section we move on a generalization of the present scheme in terms of a global analysis, i.e., we implement in the backreaction profile the radial and vertical morphology of the real background configuration.

6. Global reformulation of the model

We now generalize the scheme above in order to account for the background profile of the disk, by explicitly including its radial and vertical dependence in the equilibrium configuration. All the basic quantities involved in the equilibrium setting retains the same form as in the previous Section with the basic simplification that the poloidal velocity field and the azimuthal magnetic component are set equal to zero. For the sake of completeness, we restate the problem *ab initio* retaining, when possible, the same previously adopted notation.

We deal with an axial symmetric disk, surrounding a central astrophysical object as a star of mass M_S that is endowed with a proper magnetic field $\vec{B}$ expressed in terms of the magnetic flux functions $\psi = \psi(r, z^2)$ (see Eq.(15) with $I = 0$). Furthermore we assume that the magnetic field of the central object can be properly described by a dipole-like configuration defined as

$$\psi_D(r, z^2) = \mu_0 r^2 / (r^2 + z^2)^{3/2}, \qquad \mu_0 = \text{const}, \tag{41}$$

where the constant value μ_0 fixes the dipole field amplitude. From the Newton potential χ that describes the gravitational field generated by the central object the following Keplerian angular velocity ω_K can be defined

$$\omega_K^2(r, z^2) = GM_S / (r^2 + z^2)^{3/2}. \tag{42}$$

Because of the function form of ω_K and ψ_D, the following meaningful relations hold on the equatorial plane $z = 0$

$$\left.\begin{array}{l} \omega_K^2(r, 0) \equiv \omega_{K0}^2 = \dfrac{GM_S}{r^3} \\[2mm] \psi_D = \dfrac{\mu_0}{r} \end{array}\right\} \quad \Rightarrow \quad \omega_{K0}^2 = GM_S \psi_D^3 / \mu_0^3. \tag{43}$$

Furthermore, from the corotation theorem[13] it is natural to postulate that the relation

$$\omega^2 = GM_S \psi^3 / \mu_0^3, \tag{44}$$

holding on the equatorial plane, is valid everywhere in the disk.

Finally, we can write down the equilibrium equations describing such a configuration, which stand as

$$GM_S \epsilon \left[-\frac{r\psi^3}{\mu_0^3} + \frac{\psi_D}{\mu_0 r} \right] = -\partial_r p - \frac{\partial_r \psi}{4\pi r} \mathcal{P}[\psi], \tag{45}$$

$$-\partial_z p - \epsilon \frac{GM_S z}{\mu_0 r^2} \psi_D - \frac{\partial_z \psi}{4\pi r} \mathcal{P}[\psi] = 0, \tag{46}$$

where

$$\mathcal{P}[\psi] \equiv \partial_r \left(\frac{1}{r} \partial_r \psi \right) + \frac{1}{r} \partial_z^2 \psi. \tag{47}$$

6.1. *Perturbation equations: toward the crystalline structure*

We assume that the main contribution to the magnetic field is given by that of the central object, plus a perturbation term ζ due to the backreaction within the disk. This can be restated in terms of the flux surface function ψ as follows:

$$\psi = \psi_D + \zeta, \qquad |\zeta| \ll |\psi_D|. \tag{48}$$

This way, the equilibrium configuration equations rewrite as

$$\epsilon \frac{GM_S}{\mu_0} \left[-\frac{r}{\mu_0^2} \left(\psi_D^3 + 3\psi_D^2 \zeta \right) + \frac{\psi_D}{r} \right]$$
$$= -\partial_r p - \frac{1}{4\pi r} \left(\partial_r \psi_D + \partial_r \zeta \right) \mathcal{P}[\zeta], \tag{49a}$$

$$\partial_z p + \epsilon \frac{GM_S z}{\mu_0 r^2} \psi_D + \frac{1}{4\pi r} \left(\partial_z \psi_D + \partial_z \zeta \right) \mathcal{P}[\zeta] = 0, \tag{49b}$$

where we retained the gradients of ζ because of the hierarchical ordering due to the small scale of the backreaction.

Since the disk is assumed thin, we can approximate $(r^2 + z^2)^a \simeq r^{2a}(1 + az^2/r^2)$; Eq. (49) can be then recast as

$$\frac{3\epsilon GM_S}{r^4} \left(z^2 - \frac{(r^3 - 3rz^2)\zeta}{\mu_0} \right)$$
$$= -\partial_r p - \frac{1}{4\pi r} \left[-\frac{\mu_0}{r^2} \left(1 - \frac{9z^2}{2r^2} \right) + \partial_r \zeta \right] \mathcal{P}[\zeta], \tag{50a}$$

$$\partial_z p + \epsilon \frac{GM_S z}{r^3} + \frac{1}{4\pi r} \left(-\frac{3\mu_0 z}{r^3} + \partial_z \zeta \right) \mathcal{P}[\zeta] = 0. \tag{50b}$$

As previously adopted, we will denote background quantities by an overbar and perturbation with a hat; furthermore, we require that the quantities $\bar{\epsilon}$ and $\bar{p}$ are linked via the isothermal relation $\bar{p} = v_s^2 \bar{\epsilon}$, while we determine the form of the latter by imposing the validity of the gravothermal vertical equilibrium

$$\partial_z \bar{p} + \bar{\epsilon} \omega_{K0}^2 z = 0 \Rightarrow \bar{\epsilon} = \epsilon_0(r) \exp \left[-\omega_{K0}^2 z^2 / 2v_s^2 \right], \tag{51}$$

here $\epsilon_0(r)$ denotes the mass density on the equatorial plane. Finally, we require the following additional conditions

$$\frac{\zeta r}{\mu_0} = \frac{\zeta}{\psi_{D0}} \gg \frac{3z^2}{2r^2}, \quad \frac{z^2}{r^2} \ll 1, \quad \frac{\zeta}{\psi_{D0}} \gg \gamma \frac{rv_s^2}{3GM_S}, \tag{52}$$

$(\psi_{D0} = \psi_D(r, z = 0))$.

Let us now fix the region of validity for the crystalline micro-structure of the disk. Denoting by k the wavenumber of the radial dependence characterizing the redefined function $\phi = \zeta/\sqrt{r}$, we restrict the analysis to the disk zone where $kr \gg 1$. Thus, implementing the ansatz $\epsilon_0 \propto r^\gamma$ (where, $\gamma = $ const (see below)), and neglecting the quantity $\partial_z \psi_D$ in the vertical force balance, the disk configuration is determined via the following system:

$$\frac{3GM_S\epsilon_0(r)}{\mu_0\sqrt{r}} \left(\exp\left[-\frac{\omega_{K0}^2 z^2}{2v_s^2} \right] + \hat{D} \right) \left(1 - 3\frac{z^2}{r^2} \right) \phi$$
$$= \partial_r \hat{p} + \frac{1}{4\pi r} \left[\partial_r \phi - \frac{\mu_0}{r^{5/2}} \left(1 - \frac{9z^2}{2r^2} \right) \right] (\partial_r^2 \phi + \partial_z^2 \phi), \tag{53a}$$

$$\partial_z \hat{p} + \hat{\epsilon}\frac{GM_S z}{r^3} + \frac{1}{4\pi r}\partial_z \phi \left(\partial_r^2 \phi + \partial_z^2 \phi \right) = 0, \tag{53b}$$

where we redefined $\hat{D} = \hat{D}(r, z^2) \equiv \hat{\epsilon}/\epsilon_0$. We note how, being h the characteristic scale for the z-dependence of ζ, the possibility to neglect the term $\partial_z \psi_D$ is strictly related to the validity of the assumption $\zeta/\psi_D \gg 3zh/r^2$.

6.2. *The linear case*

As before, the linear case corresponds to a sufficiently low electromagnetic backreaction in the plasma; in this case, we neglect non-linear terms in ϕ in the equations, and we express the mass density as

$$\bar{\epsilon} \simeq \epsilon_0(r) \left(1 - \frac{GM_S z^2}{2r^3 v_s^2} \right) = \epsilon_0(r) \left[1 - \left(\frac{R_S c^2}{4r v_s^2} \right) \frac{z^2}{r^2} \right], \tag{54}$$

$R_S \equiv 2GM_S/c^2$ being the Schrwarzshild radius of the star. If we restrict to that region of the disk where the inequality

$$\frac{R_S}{4r} \gg \frac{v_s^2}{c^2}, \tag{55}$$

holds, we can approximate the configurational system (53) as

$$\epsilon_0 \frac{3GM_S}{\mu_0} \left(1 - \frac{L_s z^2}{r^3} \right) \phi + \frac{\mu_0}{4\pi r^3} \left(\partial_r^2 \phi + \partial_z^2 \phi \right) = 0, \tag{56a}$$

$$\partial_z \hat{p} + \hat{\epsilon}\frac{GM_S z}{r^3} = 0, \tag{56b}$$

where $L_s \equiv GM_S/2v_s^2$. We note how in such equations the Lorentz force dominates the radial pressure gradient, and that we can also neglect $\hat{D}$, because of $\bar{\epsilon} \gg \hat{\epsilon}$ in the linear regime. Furthermore, condition (55) reads as $r \ll L_s$, as implied by condition (52).

While Eq. (56b) shows us how the perturbations behave the same way as the background, Eq. (56a) can be restated as

$$\left(\partial_r^2 \phi + \partial_z^2 \phi\right) = -(1 - L_s z^2/r^3)k^2 \phi , \tag{57}$$

as soon as we take $\epsilon_0 = m/r^3$ ($m = $ const.) and $k^2 = 12\pi G M_S m/\mu_0^2$.

For $r \gg (L_s/k^2)^{1/3}$, such equation admits the following solution that restores, under the set of conditions we fixed above, the crystalline structure emerging in the local model

$$\phi(r, z^2) = A \sin(kr) \exp\left[-k\sqrt{L_s} z^2/2 r^{3/2}\right] , \tag{58}$$

where A is a constant amplitude. The local oscillating behavior of ζ is sketched in Fig. 1.

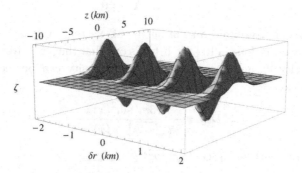

Fig. 1. Local oscillating behavior of the perturbed flux function ζ around a radius $r = 10^3$km (δr is the half-width of the considered radial region). The chosen parameters are: $A = 1$, $T = 10^7$K, $L_s \simeq 8 \times 10^5$Km, $B = 10^{12}$Gauss and $m = 0.02 M_\odot$.

It is possible to summarize in the following three physical restrictions the basic assumption on which our model relies on

(1) The disk is assumed to be thin, i.e., $z^2/r^2 \ll 1$.
(2) The wavelength of the perturbations, due to the electromagnetic back-reaction, is much smaller than the disk size, i.e., $kr \gg 1$.
(3) The disk region of the analysis must be restricted where the condition $r \ll L_s$ holds, i.e.,

$$r \ll \frac{GM_S}{2v_s^2} = \frac{R_S c^2}{4 v_s^2} = \frac{m_i c^2}{8 K_B T^2} R_S . \tag{59}$$

Thus, this scheme requires the disk temperature T not to exceed a given value for the existence of the crystalline structure.

6.3. *Extreme non-linear regime*

Let us now analyze the case when the electromagnetic backreaction domi-
nates the background dipole contribution (i.e., $\partial_r\zeta \gg \partial_r\psi_D$) and the perturbed mass density $\hat{\epsilon}$ is larger than $\bar{\epsilon}$; furthermore, we impose that the
gravitational term be negligible with respect to the pressure gradient

$$r/\hat{L}_s \gg 2hz/r^2 \,, \tag{60}$$

where $\hat{L}_s \equiv GM_S/2\hat{v}_s^2$ (with $\hat{p} = \hat{v}_s^2\hat{\epsilon}$). Fixing the mass density on the
equatorial plane in the form $\epsilon_0 = m'/\sqrt{r}$ (with $m' = $ const.), and setting
$\hat{p}(r, z^2) = \hat{q}(r, z^2)/r$, the configuration system (53) reads as

$$\frac{3GM_S m'}{\mu_0}\hat{D}\phi = \partial_r\hat{q} + \frac{1}{4\pi}\partial_r\phi\Delta\phi \,, \tag{61a}$$

$$\partial_z\hat{q} + \frac{1}{4\pi}\partial_z\phi\Delta\phi = 0. \tag{61b}$$

It is possible to obtain a solution to this system of equations in close
analogy to what is done in Ref. 8. It turns out that the (effective) perturbed
sound velocity $\hat{v}_s^2$ and the perturbed mass density can be recast in the
following form

$$\hat{\epsilon} = \hat{D}\epsilon_0 = \frac{3m'\sin^2(x)}{8\pi\sqrt{x/k}\left[\cos(x) + 4\right]} \,, \tag{62}$$

where $x = \tilde{k}r$ (with $\tilde{k} = (3GM_S m'/\mu_0^2)^{2/9}$).

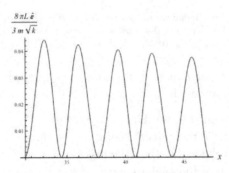

Fig. 2.   Dimensionles perturbed mass density $\hat{\epsilon}$ vs dimensionless radius x obtained in
the extreme non linear regime described above for $A = 1$, $B = 1/8$ and $F\left(u^2\right) = 1$.

The radial behavior of this perturbed density is depicted in Fig. 2. It
is worth noting how a ring-sequence decomposition of the disk, i.e., the
radial oscillation of the density, appears when the induced magnetic field

dominates the pre-existing one. Indeed, a thin disk can admit a plasma configuration characterized by micro-structures which have the form of double and opposite current filaments. It is natural to expect that the instability properties of this radial profile has a relevant role in the transport dynamics, especially in view of energy and angular momentum transfer from the micro- to the meso-scales of the system.

7. Outlooks: cross-fertilization with laboratory plasmas

It is easy to realize how the (local and global) 2D analysis here addressed be a clear reformulation of the accretion process in terms of the full MHD scenario, consistently with the axial symmetry and the quasi-ideal nature of the disk plasma. This new paradigm offers a very valuable arena to search for a fruitful cross-fertilization between the astrophysical and laboratory settings. By other words, in order to avoid the difficulties of the Shakura model (in particular, the anomalous resistivity puzzle) it is possible to export some features of the transport in laboratory plasmas to the context of a steady quasi-ideal configuration surrounding a compact astrophysical object. Here, we just sketch the basic perspective in such a direction by inferring a possible mechanism of plasma infalling even when the dissipative phenomena do not pay for the angular momentum transfer across the disk.

Indeed, the request for an anomalous resistivity is due to necessity of balancing the Lorentz force in the generalized Ohm law. Otherwise, such an Ohm equation would reduce, for a thin disk configuration (having a negligible radial magnetic component), to the constrain $v_r B_z \simeq 0$. However, this restriction for the configuration variables admits also the solution $B_z \simeq 0$ in addition to the issue $v_r \simeq 0$, which is clearly inconsistent with accretion features. The solution $B_z \simeq 0$ is clearly forbidden in the Shakura model, where the backreaction is negligible and the vertical magnetic component is almost a dipole-like contribution. Nonetheless, in the 2D reformulation, when the extreme non-linear limit takes place, the induced magnetic field inside the disk can overcome the background one, acquiring the typical radial oscillation of the crystalline profile, as we discussed above.

In such a non-linear configuration, as soon as the small scale nature of the backreaction is taken into account, it is immediate to recognize that a large number of X-points of the magnetic field, where $B_z \equiv 0$, arise and there is a finite measure region of the space where $B_z \sim 0$, allowing a non-zero infalling velocity of the plasma, even in the absence of dissipation. Such a porosity is commonly observed in laboratory plasma and it can be reliable

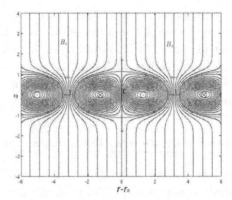

Fig. 3.   Qualitative sketch of the induced magnetic field configuration.

exported in the astrophysical scenario to account for the plasma accretion onto a compact object. Clearly, this is just a qualitative statement and it is necessary to postulate the existence of a process able to efficiently pump the plasma across the X-point, in order to account for the observed stellar accretion rates.

It has been tentatively argued (see,[14]) that the pumping mechanism can be identified in the so-colled *ballooning modes*,[15] which are unstable plasma behaviors well-localized along the symmetry axis. By other worlds, we conclude the present review of the accretion picture in axial symmetry by inferring the following scenario according to the sketch of Fig.3: in the region around the fiducial radius the magnetic field lines are crowded between two adjacent separatrices and, thus, the flow across them can be envisioned as being maintained intermittently by recurrent resistive ballooning modes that let the plasma slip through the magnetic field lines. These modes can be driven by the combined effect of the local radial density gradient and of gravity.

This work was developed within the framework of the CGW Collaboration (www.cgwcollaboration.it) and the authors would like to thank Bruno Coppi for providing Fig.3.

References

1. J.E. Pringle and M.J. Rees, *A&A* **21**, 1 (1972).
2. N.I. Shakura and R.A. Sunyaev, *A&A* **24**, 337 (1973).
3. D. Lynden-Bell and J.E. Pringle, *Mon. Not. RAS* **168**, 603 (1974).

4. R. Ruffini and J.R. Wilson, *Phys. Rev. D* **12**, 2959 (1975).
5. B. Coppi, *Phys. Plasmas* **12**, 7302 (2005).
6. N.I. Shakura, *Soviet Astro.* **16**, 756 (1973).
7. G.S. Bisnovatyi-Kogan and R.V.E. Lovelace, *New Astro. Rev.* **45**, 663 (2001).
8. B. Coppi and F. Rousseau, *ApJ* **641**, 458 (2006).
9. G. Montani and R. Benini, *Gen. Rel. Grav.* **43**, 1121 (2011).
10. G. Montani and R. Benini, *Phys. Rev. E* **84**, 026406 (2011).
11. S.A. Balbus and J.F. Hawley, *Rev. Mod. Phys.* **70**, 1 (1998).
12. S.A. Balbus, *Annual Rev. A&A* **41**(1), 555 (2003).
13. V.C.A. Ferraro, *Mon. Not. RAS* **97**, 458 (1937).
14. R. Benini and G. Montani, In *Proceedings of the 12th Marcel Grossmann Meeting (Paris - France, 2009)*, ISBN: 978-981-4374-51-4.
15. B. Coppi, *Phys. Rev. Lett.* **39**, 939 (1977).

Chapter 7

From reformulations of quantum many-body problems in- and out-of-equilibrium to applications to solar energy conversion on the nanoscale

S. Pittalis, A. Delgado, C.A. Rozzi

Institute of Nanoscience, Consiglio Nazionale delle Ricerche (Modena, Italy)

We provide a minimal introduction to density functional theory (DFT) and its time-dependent (TD) extension (TDDFT). As an example of the power of the methods, we briefly review recent results on the charge-separation dynamics potentially useful in solar energy conversion.

1. Introduction

Most of the variety and complexity of states of matter can be, in principle, described and predicted by solving the many-body time-dependent Schrödinger equation involving electrons and nuclei. However, tackling this problem directly would not bring us very far. Electrons and nuclei interacts among themselves and these many-body interactions severely challenge any analytical attempt making a closed analytical solution not even available for three interacting bodies. In addition, the number of degrees of freedoms underlying the state of the many electrons and nuclei in materials (even in single molecules) severely challenge also our present computational resources.

Hence, it is essential to simplify and reformulate the problem of the structure of matter. Typically, this is achieved by virtue of the Born-Oppenheimer (BO) approximation which allows us to decouple the electronic and nuclear degrees of freedom for systems with characteristic energies for the electronic excitations significantly larger than the energy scales of the vibrational or rotational motions. On top of this approximated picture, the problem can be further simplified by treating nuclei as classical point particles that move in an energy surface determined by solving the quantum-mechanical Schrödinger equation for the electrons in the material.

The solution of the electronic structure can be addressed from different angles. Post-Hartree-Fock wavefunction based approaches[1] have been extensively developed by the quantum chemistry community and offer sophisticated methods to capture the electronic correlations. However, all of these methods become computationally prohibited already for small molecules containing tens of atoms. Moreover, we should consider that experiments usually probe some collective or global variables rather than each single degree of freedom in the system. This suggests that we may be able to reformulate the problem in terms of some more accessible quantities rather than the full many-body wavefunction. Time-dependent density-functional theory (TDDFT) is such a reformulation in which the focus is on the electron density

$$n(\mathbf{r}, t) := N \int d\mathbf{r}_2 ... \int d\mathbf{r}_N \left| \Psi(\mathbf{r}, \mathbf{r}_2, ..., \mathbf{r}_N, t) \right|^2 ; \qquad (1)$$

here $\Psi(\mathbf{r}_1, \mathbf{r}_2, ..., \mathbf{r}_N, t)$ is an N-electron wavefunction ($\mathbf{r}$ represents the spatial position of an electron – spin degrees of freedom and the sum over all of them are understood). The particle density represents the distribution of the particles and it integrates to their total number. Similarly DFT focus on the ground state electron density $n(\mathbf{r})$. We shall before review the basic ideas of DFT and then consider how these are extended to the time-dependent case to tackle excitations and dynamics.

Our objective is to illustrate some introductory concepts in a reader-friendly fashion. It is not our aim to provide a review of the status of the field. This should justify a largely incomplete list of references. An interested reader eager to learn more will find it useful to study on advanced textbooks. For DFT, we suggest the book by Dreizler and Gross,[2] the more recent book by Engel and Dreizler,[3] and the textbook by Parr and Yang.[4] For TDDFT, we suggest the book by Ullrich[5] and the volume of lecture notes from the TDDFT School of Benasque.[6] Finally we point out the textbook by Giuliani and Vignale which covers DFT, TDDFT, many-body perturbation theory with an emphasis on on the uniform electron gas.[7]

2. Density Functional Theory

If the system is unperturbed, the electronic Hamiltonian is time independent,

$$\hat{H}_0 = \hat{T} + \hat{V}_{ee} + \hat{V}_0 , \qquad (2)$$

where (in atomic units such that $\hbar = m = e = 1$) $\hat{T} = \sum_i -\frac{\nabla_i^2}{2}$ is the electrons' kinetic operator, $\hat{V}_{ee} = \frac{1}{2}\sum_{i \neq j} \frac{1}{|\mathbf{r}_i - \mathbf{r}_j|}$ is the electron-electron interaction, and $\hat{V}_0$ represents the scalar potential felt by the electrons due to the Coulomb attraction with the nuclei and, possibly, to an additional external *static* potential. With clamped nuclei at the equilibrium positions, the nuclear-nuclear interaction contributes as an overall constant that may be disregarded.

Specified the number of electrons and the form of their mutual interaction, $\hat{H}_0$ is determined by specializing the last term. Given $\hat{V}_0(\mathbf{r}_i)$, the total energy can be viewed as a functional of the many-body wavefunction:

$$E_{V_0}[\Psi] = \langle \Psi | \hat{T} + \hat{V}_{ee} + \hat{V}_0 | \Psi \rangle . \tag{3}$$

At standard low temperatures, we are particularly interested in ground state properties. The functional in Eq. (3) may be evaluated for any N-electron wavefunction, and the Rayleigh-Ritz variational principle ensures that the ground state energy, E_{V_0}, is given by

$$E_{V_0} = \inf_{\Psi} E_{V_0}[\Psi], \tag{4}$$

where the infimum is taken over all normalized, antisymmetric wavefunctions [a]. Choosing the form of the many-body wavefunctions, the Rayleigh-Ritz variational principle provides us a procedure for finding approximate solutions. Unfortunately, this approach is afflicted by an impractical growth of the numerical effort with the number of particles.

Inspired by the Thomas-Fermi (TF) approach,[8-10] one might wonder if the role played by the wavefunction could be played by the particle density. In that case, one would deal with a function of only three spatial coordinates, regardless of the number of electrons. The TF theory is an approximate approach later rigorously justified, through a semiclassical analysis, by Lieb and Simon for all the non-relativistic matter with atomic number growing to infinity.[11] However, about ten years earlier the work by Lieb and Simon, Hohenberg and Kohn proved that the idea to use solely the particle density as cornerstone for the many-body problem is in principle correct. The Hohenberg-Kohn (HK) theorem[12] connects specific sets of densities, wavefunctions, and potentials in such a way that it assures us that the electronic density alone is enough to determine *all* observable quantities of the systems.

---

[a]We also assume that some boundary spatial conditions have been specified for the admissible wavefunctions. Parametric dependence of all the quantities on the nuclear position is understood.

For the sake of simplicity, here we restrict ourselves to systems with *non* degenerate ground states with wavefunctions which go to zero fast enough at infinity (it is understood that proper generalizations are possible). Let $\mathcal{P}$ be the set of all external potentials leading to a reasonable ground state for N electrons. Two of these potentials are considered to be different if they differ more than an additive constant. For a given potential, the corresponding ground state, Ψ, is obtained through the solution of the Schrödinger equation. Wavefunctions obtained in this way form the set $\mathcal{W}$ of the interacting v-representable wavefunctions. The corresponding particle densities can be computed using definition (1). Ground state particle densities so obtained form the set of the interacting v-representable densities $\mathcal{D}$. *The HK theorem proves that the mappings above are invertible as well.* Therefore, the theorem concludes that the ground state particle density ultimately determines the external potential from which it comes: i.e., the potential (modulo a trivial constant) is a functional of the particle density: $v = v[n]$. This means that, in principle, through the ground state particle density we can determine the Hamiltonian – for a given interaction and particle number – and derived quantities!

In particular, we see that, all the ground-state observables can be expressed as density functional because the ground state wavefunction is a functional of the ground state density $\Psi = \Psi[n]$ (for this it is sufficient to exploit the one-to-one correspondence between $\mathcal{W}$ and $\mathcal{D}$). Thus, Hohenberg and Kohn define the density energy functional

$$E_{v_0,\mathrm{HK}}[n] := F_{\mathrm{HK}}[n] + \int d^3 r \, n(\mathbf{r}) v_0(\mathbf{r}) \,, \tag{5}$$

where $\hat{V}_0 = \sum_{i=1}^{N} v_0(\mathbf{r}_i)$, $F_{\mathrm{HK}}[n] := \langle \Psi[n] | \hat{T} + \hat{V}_{ee} | \Psi[n] \rangle$, and $n(\mathbf{r})$ is any density in $\mathcal{D}$. Note, $F_{\mathrm{HK}}[n]$ is the same for *all* the systems represented by the potentials in $\mathcal{P}$. In this sense, $F_{\mathrm{HK}}[n]$ is a *universal* density functional!

Let n_0 be the ground-state particle density of the potential v_0. The Rayleigh-Ritz variational principle (4) immediately tells us

$$E_{v_0} = \min_{n \in \mathbf{D}} E_{v_0,\mathrm{HK}}[n] = E_{v_0,\mathrm{HK}}[n_0] \,. \tag{6}$$

Hence, a variational principle based on the particle density (instead of the computationally expensive wavefunction) is available.

We may now search for an approximation to $F_{\mathrm{HK}}[n]$ which could be applicable to many different problems and require only a minimal computational burden. But balance with accuracy demands the consideration of an auxiliary non-interacting problem; i.e., the Kohn-Sham (KS) system.[13]

For this, let us consider *non-interacting* electrons acted by some external local potentials $v_s(\mathbf{r})$. We can mirror the procedure applied to the interacting electrons and group the possible $v_s(\mathbf{r})$ in the set $\mathcal{P}_s$. The corresponding non-interacting ground state wavefunctions Φ are then grouped in the set $\mathcal{W}_s$. Finally, their particle densities n_s form the set $\mathcal{D}_s$. We can then apply the HK theorem and define the non-interacting analog of F_{HK}, which is simply the kinetic energy

$$T_s[n_s] = \langle \Phi[n_s] | \hat{T} | \Phi[n_s] \rangle .   \tag{7}$$

At this point, Kohn and Sham assumed that for each element n of $\mathcal{D}$, a potential v_s in $\mathcal{P}_s$ exists, with corresponding ground-state particle density $n_s = n$. v_s is the KS potential. In other words, interacting v-representable densities are also assumed to be non-interacting v-representable. This maps the interacting problem onto a non-interacting one [b]. Since we continue to consider only non-degenerate ground states, the KS ground-state wavefunction is a single Slater determinant. Note that the KS orbitals are proper Fermionic single-particle states. Therefore, the ground-state KS wavefunction is obtained by occupying the eigenstates with lowest eigenvalues obtained form the solution of the non-interacting problem

$$\left[-\frac{1}{2}\nabla^2 + v_s(\mathbf{r}) \right] \varphi_i(\mathbf{r}) = \epsilon_i \varphi_i(\mathbf{r}) .   \tag{8}$$

Thus, the corresponding density is readily obtained

$$n(\mathbf{r}) = \sum_{i=1}^{N/2} n_i |\varphi_i(\mathbf{r})|^2,   \tag{9}$$

where n_i is the occupation number of the ith single particle state (this also accounts for spin degeneracy).

The usefulness of the KS scheme stems from the fact that it allows us to express $F_{HK}[n]$ in terms of the KS kinetic energy, $T_s[n_s]$, the Hartree electrostatic energy [c] $E_H[n] = \frac{1}{2} \int \int d^3r d^3r' \frac{n(\mathbf{r})n(\mathbf{r}')}{|\mathbf{r}-\mathbf{r}'|}$, and a reminder, $E_{xc}[n] := F_{HK}[n] - T_s[n_s] - E_H[n]$, which *defines* the exchange-correlation (xc) energy. Thus, we have:

$$F_{HK}[n] = T_s[n] + E_H[n] + E_{xc}[n] .   \tag{10}$$

---

[b]We refer discussions of interacting and non-interacting v-representability conditions to the literature suggested in the introduction.
[c]The Hartree electrostatic energy is introduced as a useful starting point for an explicit (classical) estimation of the electron-electron interaction energy.

Now we can also give a formal expression for the KS potential. Assuming differentiability of all the functionals, the necessary conditions for having energy minima for both the interacting and KS system translate into

$$v_s(\mathbf{r}) = v_0(\mathbf{r}) + \int d^3 r' \frac{n(\mathbf{r}')}{|\mathbf{r} - \mathbf{r}'|} + v_{\mathrm{xc}}[n](\mathbf{r}) \qquad (11)$$

where the xc potential is given by $v_{\mathrm{xc}}[n](\mathbf{r}) = \frac{\delta E_{\mathrm{xc}}[n]}{\delta n(\mathbf{r})}$.

Though often small, the $E_{\mathrm{xc}}[n]$ still represents an essential part of the total energy. Its exact form is unknown, and it therefore must be approximated in practice. Good and surprisingly efficient approximations exist for $E_{\mathrm{xc}}[n]$. The first and simplest approximation to $E_{\mathrm{xc}}[n]$ is the Local Density Approximation (LDA).[13] The LDA assumes that the xc energy density can be approximated locally with that of the uniform gas. The LDA can be taken as a base on which to build on to reach better accuracies.[14] The concept of the exchange-correlation hole has been vital to derive important approximations.[17,18] Inclusion of density gradient dependence – in the fashion of the Generalized Gradient Approximations (GGAs) – generates sufficiently accurate results to be useful in many chemical and materials applications.[16–18] Hybrid functionals substitute a fraction of single-determinant exchange for part of the GGA exchange;[19–21] meta-GGAs[14,15] introduce a dependence on the kinetic energy density; and hyper-GGAs[14,15] include exact exchange as input to the functional. Functional forms can be derived from many-body perturbation theory, in this case the dependency on the density becomes increasingly more implicit and inclusion of unoccupied orbitals as inputs increase complexity and computational cost.

3. Time-Dependent DFT

The analogous theorem which plays the role of the HK theorem in DFT for or the time-dependent case was proven by Runge and Gross (RG) in 1984.[22] They showed that if two N-electron systems start from the same initial state, but are subject to two different time-dependent potentials, their respective time-dependent densities will be different. Under the restrictions stated in the previous section and for potentials with Taylor expansion about the initial time, the RG theorem implies that

$$v(\mathbf{r}, t) + c(t) \xleftrightarrow[\text{(fixed } \Psi_0)]{\text{one-to-one}} n(\mathbf{r}, t) . \qquad (12)$$

This allows us to rewrite the potential as a functional of the $n(\mathbf{r})$ and Ψ_0:

$$v(\mathbf{r}, t) = v[n, \Psi_0](\mathbf{r}, t) . \qquad (13)$$

As a consequence, the time-dependent Hamiltonian exhibits such a functional dependence. Thus, the corresponding time-dependent wavefunction and observables are also functionals of $n(\mathbf{r}, t)$ and Ψ_0.

Now let us assume that the density $n(\mathbf{r}, t)$ which evolves in an interacting system under the action of an external potential $v(\mathbf{r}, t)$, starting from an initial state Ψ_0, can also be reproduced in a non-interacting system [d]. Under the action of an appropriate and uniquely determined potential $v_s(\mathbf{r}, t)$, starting from an initial state Φ_0. We require that Φ_0 has the same density and divergence of the current density as Ψ_0. Therefore, we can rewrite

$$n(\mathbf{r}, t) = \sum_{i=1}^{N} n_i |\varphi_i(\mathbf{r}, t)|^2 \tag{14}$$

where $\varphi_i(\mathbf{r}, t)$ satisfy the time-dependent Kohn-Sham equation:

$$i\frac{\partial}{\partial t}\varphi_j(\mathbf{r}, t) = \left[-\frac{\nabla^2}{2} + v_s(\mathbf{r}, t) \right] \varphi_j(\mathbf{r}, t), \tag{15}$$

with an effective time-dependent potential given by

$$v_s(\mathbf{r}, t) = v(\mathbf{r}, t) + \int d^3 r' \frac{n(\mathbf{r}', t)}{|\mathbf{r} - \mathbf{r}'|} + v_{\mathrm{xc}}[n, \Psi_0, \Phi_0](\mathbf{r}, t). \tag{16}$$

Note that the occupation numbers, n_i in Eq. (14) do not exhibit any time evolution, though the single particle orbitals occupied at the initial time change in time. The time-dependent xc potential, $v_{\mathrm{xc}}(\mathbf{r},t)$ has a functional dependence on the density, the initial many-body state Ψ_0 of the interacting system, and the initial state of the KS system Φ_0. The dependence on the initial states drop outs if the system starts from the ground state; simply because the HK theorem tells us that a ground-state wavefunctions are functional of the ground-state densities.

We should emphasize that the potentials v and v_s at time T are functionals of the *history* of the densities over the whole time interval $[t_0, T]$. Proper consideration of the dependence on the initial state and density history provides insights on exact properties of the TDDFT xc potential.[23]

In close analogy with static DFT, where the potentials are functional derivatives of energy functionals with respect to the density, the potentials at time T can be generated from functional derivative of an action functional with respect to the density at the same time. The existence of such a representation was suggested by RG in their original paper.[22] Notice that

---

[d]We refer discussions of time-dependent interacting and non-interacting v-representability conditions to the literature suggested in the introduction.

(ignoring the dependence on the initial state, which is irrelevant in this context)

$$\mathcal{A}_v[n] = \int_0^T dt \; \langle \psi[n]|i\partial_t - \hat{H}|\psi[n]\rangle = \mathcal{F}[n] - \int_0^T dt \int d\mathbf{r} \; n(\mathbf{r},t)v(\mathbf{r},t) \;,$$

(17)

where

$$\mathcal{F}[n] \equiv \int_0^T dt \; \langle \psi[n]|i\partial_t - \hat{T} - \hat{V}_{ee}|\psi[n]\rangle$$

(18)

is a universal functional of the density. It was realized that this representation was plagued by the so-called *the causality paradox*.[24] Sophisticated solutions[25–28] were proposed, till Vignale provided a straightforward and (pedagogically) transparent resolution in 2008.[29]

The key observation is that the Frenkel variational principle $\delta A_V = 0$ is valid only for variations of Ψ that vanish at the endpoints of the time interval under consideration, i.e. at $t = 0$ and $t = T$. But a variation of the density at any time $t < T$ inevitably causes a change in the quantum state at time T. Therefore, the correct formulation of the variational principle for the density is not $\delta A_v = 0$ but

$$\delta\mathcal{A}_v[n] = i\langle \Psi_T[n]|\delta\Psi_T[n]\rangle \;.$$

(19)

Here $\Psi_T[n] \equiv \Psi[n](T)$ is the quantum state at time T regarded as a functional of the density. It is apparent that the action functional is not stationary but its variation must be equal to another functional of the density [the right hand side of Eq. (19)]. Taking the functional derivative of Eq. (19) with respect to $n(\mathbf{r},t)$ and making use of Eq. (17), we get

$$v(\mathbf{r},t) = \frac{\delta\mathcal{F}[n]}{\delta n(\mathbf{r},t)} - i\left\langle \Psi_T[n] \left| \frac{\delta\Psi_T[n]}{\delta n(\mathbf{r},t)} \right. \right\rangle \;,$$

(20)

where $\left| \frac{\delta\psi_T[n]}{\delta n(\mathbf{r},t)} \right\rangle$ is a compact representation for the functional derivative of the *ket* $|\psi_T[n]\rangle$ with respect to density. Both terms in Eq. (20) exhibit the causality problem if taken alone, but the overall effects cancels out and the right hand side does not depend on the later time T. Similarly we can proceed for the KS action functional and obtain

$$v_s(\mathbf{r},t) = \frac{\delta\mathcal{F}_s[n]}{\delta n(\mathbf{r},t)} - i\left\langle \Phi_T[n] \left| \frac{\delta\Phi_T[n]}{\delta n(\mathbf{r},t)} \right. \right\rangle \;,$$

(21)

where

$$\mathcal{F}_s[n] = \int_{t_0}^T dt \langle \Phi[n](t)|i\frac{\partial}{\partial t} - \hat{T}|\Phi[n](t)\rangle \;.$$

(22)

Mimicking what was done in DFT for the interacting energy functional, we can decompose the interacting action functional in terms of other functionals making use of the TD KS system

$$\mathcal{F}[n] = \mathcal{F}_s[n] - \mathcal{A}_H[n] - \mathcal{A}_{xc}[n] \tag{23}$$

where $\mathcal{A}_H[n] := \int_{t_0}^{T} dt \int d^3r \int d^3r' \frac{n(\mathbf{r},t)n(\mathbf{r}',t)}{|\mathbf{r}-\mathbf{r}'|}$. Finally, one gets[29]

$$v_{xc}[n;\mathbf{r},t] = \frac{\delta \mathcal{A}_{xc}[n]}{\delta n(\mathbf{r},t)} + i \left\langle \psi_{t+}[n] \left| \frac{\delta \psi_{t+}[n]}{\delta n(\mathbf{r},t)} \right\rangle - i \left\langle \Phi_{t+}[n] \left| \frac{\delta \Phi_{t+}[n]}{\delta n(\mathbf{r},t)} \right\rangle . \tag{24}$$

A very popular approximation for $\mathcal{A}_{xc}$ reuses the approximation available for E_{xc} as follows

$$\mathcal{A}_{xc}[n] \rightarrow \mathcal{A}_{xc}^{Adiab}[n] = \int_{t_0}^{T} dt \, E_{xc}[n_{GS}]\Big|_{n_{GS}(\mathbf{r})\rightarrow n(\mathbf{r},t)} \tag{25}$$

from which

$$v_{xc}[n] \rightarrow v_{xc}^{Adiab}[n;\mathbf{r},t] = \frac{\delta \mathcal{A}_{xc}^{Adiab}[n]}{\delta n(\mathbf{r},t)} = \frac{\delta E_{xc}[n_{GS}]}{\delta n_{GS}(\mathbf{r})}\Big|_{n_{GS}(\mathbf{r})\rightarrow n(\mathbf{r},t)} \tag{26}$$

here GS stands ground state. This is the so-called *adiabatic* approximation as it becomes exact in the ideal situation in which a system evolves so slowly to stick to its *instantaneous* ground state, satisfying the adiabatic theorem of quantum mechanics.

The adiabatic approximation is often sufficient to extract precise information on the low laying excitations of finite systems such as atoms and molecules. As another case, plasmons in metallic systems are very well-described at the level of adiabatic LDA (ALDA) – because this is already captured by the response function of the KS system which includes the Hartree contribution. Dynamical XC effects beyond ALDA are essential to capture intrinsic plasmon damping caused by multiple particle-hole excitations. Non-adiabatic effects are also important for other excitations, like in charge transfers, in which the electrons move between spatially separated regions involving multi-particle excitations. Such processes can occur in complexes of two or more molecules or between different functional groups within the same molecule. Practical and efficient non-adiabatic approximations have not been yet fully established. Attempts to overcome this limitation belong to the forefront of research in TDDFT.

TDDFT applications are mostly done within the linear response regime. The literature of real-time propagation of the TDDFT equations is also growing thanks to the fact that it allows to explore the non-linear regime and requires only to deal with occupied orbitals though one must take into account the cost of the propagation in time. We summarize some recent and advanced (TD)DFT applications in the next section.

4. Charge Separation on the Nanoscale of Solar Energy Devices

In solar cells, Sun light generates neutral excitations consisting of an electron-hole in a bound state. Subsequent charge separation at the interface between an electron donor and acceptor generates charges that must be transported to the device electrodes prior to recombination. The latest generation of solar devices is mainly represented by organic and hybrid compounds. The microscopic description of the operation of such solar cells rests on the study of the alignment of energy levels of the donor and acceptor units with respect to each other and to the Fermi level of the ohmic contacts.[30] Regarding specifically the charge transfer problem within DFT, it is common practice to describe it in terms of Kohn-Sham levels and their generalizations.[31] The present state-of-the-art in modeling energy conversion in photovoltaic cell consists in analyzing the possible static configurations of a system and then in inferring the preferred channels for charge and energy transfer from ground-state energetic considerations alone. Beyond the static picture, TDDFT provides a rigorously grounded way to approach the problem of (neutral) excited-states dynamics.

As a further complication, the strong electron-phonon coupling in many organic systems[32] makes some of the most common simplifying approximation (like Born-Oppenheimer separation) not acceptable when changes in the electronic many-body wavefunction must be correlated with changes in the ionic configuration. The Ehrenfest path approximation tackles this difficulty to some extent. Here, the nuclei are evolving as classical particles under the action of the average field caused by the (quantum-mechanical) electrons and, in turn, the electrons are acted upon by the Kohn-Sham potential which includes the time-dependent potential generated by the same moving nuclei.[33] As long as the nuclear potential energy landscape does not cross conical intersections, the Ehrenfest dynamics together with TDDFT electron propagation is expected to provide a good compromise between accuracy and computational efficiency. Next, we summarized results from recent simulations[34–36] performed (mainly) by the OCTOPUS code[e].

It was already known that systems such as P3HT:PCBM bulk hetero-

---

[e]The OCTOPUS code is one of the most advanced tools for TDDFT applications in real time.[37] The code is freely available at http://www.tddft.org/

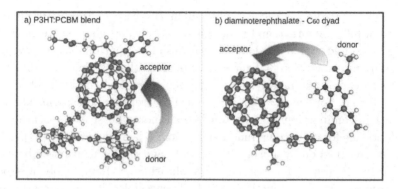

Fig. 1. (Color online) Left panel shows the structure investigated in Ref. 35; Right panel shows the structure investigated in Ref. 36. The color code for the elements is red for Oxygen, blue for Nitrogen, grey for Carbon, yellow for Sulfur, and white for Hydrogen. Structures were optimized using the B3LYP approximation for the xc energy functional.[40] The arrows represent schematically the flow of charge from the donor to the acceptor: an electron goes to the acceptor and a hole stays on the donor.

junctions are capable of an ultrafast electron and hole separation.[38] Still very little was known about the role of quantum coherence at room temperature in the earliest stage of the dynamics. Therefore the necessity to focus on the exciton dynamics of the blend in the first few hundreds of femtoseconds (fs) after photoexcitation. An initial instantaneously excited electronic state – corresponding to the removal of an electron from the polymer highest occupied DFT Kohn-Sham molecular orbital (HOMO) and the creation of an electron in the polymer lowest unoccupied DFT molecular orbital (LUMO) – mimiks the effects of the an instantaneous optical excitation due to the Sun light in the visible range. The initial conditions for the nuclear motion are sampled from a Maxwellian distribution of random nuclear velocities such to approximate the experimental room temperature environment. The TDDFT+Ehrenfest propagation revealed that an electron is transferred from the donor to the acceptor with 60% probability within 97 fs (see Fig. 1a). Moreover the charge-transfer probability oscillated in time with a period of about 25 fs, which matches approximately the oscillation frequency observed in the experiments. Remarkably, the electronic excitation remains fully localized on the polymer when keeping the ions in fixed positions. These simulations reveals that vibronic coupling is necessary for charge transfer to occur and indicate that this coupling is responsible for dynamically driving donor and acceptor in resonance, explaining the coherent oscillations of the transferred charge.[35]

Exploiting the knowledge acquired in the case of the P3HT:PCBM blend, it followed an attempt to design an entirely new photoactive molecule with a two-fold goal in mind: 1) to create a simple, controllable environment to study ultra-fast charge dynamics; 2) to allow the maximum flexibility in order to chemically modify the sample molecule, and check the effects of each change. For this, we took a combined theoretical and experimental approach. Diaminoterephthalates are colored molecular scaffolds bearing up to four sites for orthogonal functionalization. Their fluorescence properties can be switched on by a chemical reaction, and they can therefore be used as "turn-on probes" for biochemical applications.[39] Remarkably, a strong electron acceptor – a functionalized Fullerene molecule – can be attached to the original scaffold by cycloaddition reaction. In this configuration, the molecule potentially acts as an electron donor-acceptor couple. With respect to the previous case, the donor and the acceptor are chemically bound through a molecular "bridge" (see Fig. 1).

The simulation of the charge dynamical excitation[36] started from a state obtained by promoting an electron from the KS HOMO to the KS LUMO in the original KS molecular ground state. As in the previous case, this emulates an instantaneous optical excitation induced by the Sun light which mostly stays localized on the donor. The nuclei were given random velocities corresponding to a temperature of 300 K. Finally, the TDDFT+Ehrenfest propagation was initiated and the charge motion was tracked by numerical integration of the particle density in a volume containing the acceptor moiety. The same evolution was also computed by suppressing the electron-nuclei coupling by leaving the nuclei fixed at their initial equilibrium positions. Charge separation was indeed observed: an electron can move on the acceptor within 80 fs (see Fig. 1b). As in the previous case, the charge-separation was completely frozen for clamped nuclei. Of course, the road from here to an actual device is long and must include steps which require a joint multidisciplinary efforts.

Acknowledgement

We acknowledge financial support from the EU FP7 project, CRONOS (grant No. 280879-2), and the PRACE Infrastructure for access to super-computer resources at CINECA, Bologna, Italy (project No. 1562, LAIT).

References

1. J. Frank, *Introduction to computational chemistry*, John Wiley & Sons (2013).
2. R.M. Dreizler, E.K.U. Gross, *Density Functional Theory: An Approach to the Quantum Many-Body Problem* Springer–Verlag (1990).
3. E. Engel, R.M. Dreizler, *Density Functional Theory: An Advanced Course*, Springer, Berlin (2011).
4. R.G. Parr, W. Yang, *Density Functional Theory of Atoms and Molecules*, Oxford University Press (1989).
5. C.A. Ullrich, *Time Dependent Density Functional Theory: Concept and Applications*, Oxford University Press (2012).
6. *Fundamentals of Time-Dependent Density Functional Theory*, Eds. M.A.L. Marques, N.T. Maitra, F.M.S. Nogueira, E.K.U. Gross, A. Rubio, Lecture Notes in Physics, Vol. **837**, Springer–Verlag (2012).
7. G.F. Giuliani and G. Vignale, *Quantum Theory of the Electron Liquid*, Cambridge University Press (2005)
8. L.H. Thomas, *Math. Proc. Camb. Phil. Soc.* **23**(05), 542 (1927).
9. E. Fermi, *Rend. Acc. Naz. Lincei* **6**, 602, (1927).
10. E. Fermi, *Zeitschrift für Physik A Hadrons and Nuclei* **48**, 73 (1928).
11. E.H. Lieb, B. Simon, *Phys. Rev. Lett.* **83**, 681 (1973).
12. P. Hohenberg, W. Kohn, *Phys. Rev.* **136**, B864 (1964).
13. W. Kohn, L.J. Sham, *Phys. Rev.* **140**, A1133 (1965).
14. J.P. Perdew, K. Schmidt, in *Density Functional Theory and Its Applications to Materials*, ed. by V.E.V. Doren, K.V. Alsenoy, P. Geerlings (American Institute of Physics, Melville, NY, 2001).
15. A.D. Becke, *J. Chem. Phys.* 140, 18A301 (2014).
16. J.P. Perdew, *Phys. Rev. B* **33**, 8822 (1986).
17. J.P. Perdew, K. Burke, M. Ernzerhof, *Phys. Rev. Lett.* **77**(18), 3865 (1996), *ibid.* **78**, 1396(E) (1997).
18. J.P. Perdew, K. Burke, M. Ernzerhof, *Phys. Rev. Lett.* **80**, 891 (1998).
19. A.D. Becke, *Phys. Rev. A* **38**(6), 3098 (1988).
20. A.D. Becke, *J. Chem. Phys.* **98**(7), 5648 (1993).
21. C. Lee, W. Yang, R.G. Parr, *Phys. Rev. B* **37**(2), 785 (1988).
22. E. Runge and E.K.U. Gross, *Phys. Rev. Lett.* **52**, 997 (1984).
23. N.T. Maitra, in *Time-Dependent Density Functional Theory*, Lecture Notes in Physics, Vol. **706**, p. 61, Springer, Berlin.
24. E.K.U. Gross, J.F. Dobson, and M. Petersilka, in *Density Functional Theory II*, ed. by R.F. Nalewajski, Vol. **181** of Topics in Current Chemistry, p. 81, Springer, Berlin (1996).
25. A.K. Rajagopal, *Phys. Rev. A* **54**, 3916 (1996).
26. R. van Leeuwen, *Phys. Rev. Lett.* **80**, 1280 (1998).
27. R. van Leeuwen, *Int. J. Mod. Phys. A* **15**, 1969 (2001).
28. S. Mukamel, *Phys. Rev. A* **71**, 024503 (2005).
29. G. Vignale, *Phys. Rev. A* **77**, 062511, (2008).
30. S.-S. Sun and N.S. Sariciftci (Eds.), *Organic Photovoltaics, Mechanisms, Ma-*

terials and Devices, Taylor & Francis (2005).

31. A. Seidel, A. Görling, P. Vogel, J.A. Majewski, and M. Levy, *Phys. Rev. B* **53**, 3764 (1996).
32. J.L. Bredas and G. P. Street, *Acc. Chem. Res.* **18**, 309 (1985).
33. J.L. Alonso, et al., *Phys. Rev. Lett.* **101**, 096403 (2008); X. Andrade, et al., *J. Chem. Theory Comput.* **5**, 728 (2009).
34. C.A. Rozzi, et al., *Nat. Commun.* **4**, 1602 (2013).
35. S.M. Falke, et al., *Science* **344**, 1001 (2014).
36. S. Pittalis, et al., *Adv. Func. Mat.*, DOI: dx.doi.org/10.1002/adfm.201402316 (2014).
37. A. Castro, et al., *Phys. Stat. Sol. (b)* **243**, 2465 (2006).
38. G. Yu, J. Gao, J.C. Hummelen, F. Wudl, A.J. Heeger,*Science* **270**, 1789 (1995); C.J. Brabec et al., *Adv. Mater.* **22**, 3839 (2010).
39. R. Pflantz, J. Christoffers, *Chem. Eur. J.* **15**, 2200 (2009).
40. P.J. Stephens, F.J. Devlin, C.F. Chabalowski, M.J. Frisch, *J. Phys. Chem.* **98**, 11623 (1994).

Chapter 8

The NLS on a torus

C. Procesi, M. Procesi

Dipartimento di Matematica, "Sapienza" University of Rome (Rome, Italy)

Non linear PDE's are ubiquitous and among them the Non Linear Schrödinger equation, NLS for short, is a particularly instructive example due to its particular combinatorics, specially when we study it on a d–dimensional torus where the Laplace operator has positive integer eigenvalues with multiplicities which can be well understood.

This work is supported by ERC project *HamPDEs*, under FP7.

1. Introduction

1.1. *The normal form*

This chapter is devoted to discuss the content of four papers of the present authors which construct reducible quasi–periodic solutions for the completely resonant NLS equations on a d–dimensional torus $\mathbb{T}^d$:

$$iu_t - \Delta u = \kappa |u|^{2q} u + \partial_{\bar{u}} G(|u|^2). \tag{1}$$

Here $u := u(t, \varphi)$, $\varphi \in \mathbb{T}^d$, Δ is the Laplace operator and $G(a)$ is a real analytic function whose Taylor series starts from degree $q+2$, $q \geq 1$. Finally κ is a coupling constant which we can normalize to 1. This equation can be studied on any manifold in which the Laplace operator is defined, typically a Riemannian manifold. One expects different behaviour in the compact and non compact case, but even in the compact case the difficulty of the problem increases with the complexity of the eigenvalues and eigenfunctions of the Laplace operator, see[4] and[5] . The case of the torus in this sense is the simplest but nevertheless the Theory is quite complex. Only when both $d = q = 1$ one has that the NLS is completely integrable, and an extensive theory is available, our goal instead is to treat the non–integrable cases.

The first author was introduced to this topic by Michela Procesi who had a plan to deal with this type of questions using KAM theory. For the success of this plan one needs to handle, besides difficult analytic questions, also difficult combinatorial and algebraic questions where an algebraist can be of some help.

Our results are obtained by exploiting the Hamiltonian structure of equation (1) and applying an infinite–dimensional KAM algorithm. On this subject there is an extensive literature, for example we suggest,[10],[12],[7],[12] and the references therein.

As is well known such algorithms require strong *non-degeneracy* conditions not always valid, even for finite dimensional systems and, when valid, generally proved by performing on the Hamiltonian a few steps of Birkhoff normal form. This is an algebraic algorithm which in a generic case formally may transform a given Hamiltonian into a completely integrable one. But then the price is that the formal construction is everywhere divergent!

KAM theory remedies to this by showing that under suitable conditions there is some rather fractal preservation of families of quasi–periodic solutions, on the other hand the very unstable nature of non completely integrable dynamical systems implies that there is also a complementary fractal structure where the behaviour may be more cahotic. The goal is to prove that the smaller the perturbation the smaller is the part where this cahotic behaviour happens.

Of course in an infinite dimensional system quasi–periodic solutions are not typical, so one usually takes the simplified point of view to start from some large family of solutions, but depending only on finitely many frequencies, which in the end produce finite dimensional invariant tori.

A typical step of both algorithms consists in choosing a suitable Hamiltonian F, generating from F the corresponding 1–parameter group of symplectic transformations, and transforming the given Hamiltonian H into one in the new coordinates which one hopes to be simplified in some way as to make, at the end of the algorithm, the Hamiltonian vector field tangent to a constructed family of tori.

As in the classical Newton type of algorithms F is usually found by solving some linear equations called the *homological equations*. In this procedure one clearly sees that when the determinant of the system is very small this may create divergences in the algorithm and this is one of the main issues to control. However it is well known that these *small divisors* occur no matter how small the perturbation! In order to control them one then introduces parameters for a family of initial data and stepwise excizes a small part where the divisors are too small (see §3).

Imposing this excition constraints is usually called the *first Melnikov condition*. Notice that in general this linear equation is infinite dimensional so the norm in which one has to perform the estimates is quite delicate, see for instance,[6],[3].[4] In our setting though one of the first point is to show that one can start, by suitable choice of coordinates, by an infinite dimensional system which has a block diagonal form with finite dimensional blocks. The initial data can be further constrained so that this shape persists at each step of the algorithm, this is called the *second Melnikov condition* and it is verified by controlling the difference of the distinct eigenvalues in these infinitely many blocks.

In our work on the NLS, the construction of quasi–periodic solutions is performed in four steps, corresponding to 4 papers, the second also with Nguyen Bich Van.

(1) Construction of integrable normal forms, *rectangle graph*[13]
(2) Proof of non–degeneracy of the normal form, for $q = 1$, *lots of algebra*[16]
(3) Quasi–Töpliz property and the KAM algorithm, $q = 1$ *hard analysis*[14]
(4) The general case $d > 1$, $q > 1$ *algebra and analysis*[15].

The normal form is a suitable quadratic Hamiltonian depending on parameters corresponding to families of orbits of the unperturbed system, which is the *leading part* of the NLS Hamiltonian which is the sum of this plus a small perturbation. As any quadratic Hamiltonian it can be interpreted as a linear operator on a symplectic space and in our case it decouples into a direct sum of infinitely many finite dimensional blocks which can be combinatorially described and whose eigenvalues can be controlled. The construction of this form is done in[13] in which we exhibit and study the normal forms for classes of completely resonant NLS.

In the case of the cubic NLS, i.e. $q = 1$, but for all d, we have proved in[14] the existence of families of stable and unstable quasi-periodic solutions, this required a very subtle combinatorial analysis (performed in[16]) of the combinatorial blocks. This enabled us to prove the *second Melnikov conditions* (which amounts to proving that the NLS equation linearised at an approximate solution has distinct eigenvalues on the space of quasi-periodic functions). This combinatorics is not available for $q > 1$ except in dimension $d \leq 2$ (see[11]). In this context we mention the papers[9] on the cubic NLS in dimension two and the preprint.[17]

In[15] we prove that, for any d and for any value $q \in \mathbb{N}$, generically the multiplicity of the eigenvalues is uniformly bounded, and moreover there is some weak form of non–degeneracy, this leads also to existence and reducibility of quasi-periodic solutions.

1.1.1. *The Hamiltonian*

On a torus we can pass to the Fourier representation

$$u(t, \varphi) := \sum_{k \in \mathbb{Z}^d} u_k(t) e^{i(k, \varphi)} \tag{2}$$

we have, up to a rescaling of u and of time, in coordinates the Hamiltonian:

$$H := \sum_{k \in \mathbb{Z}^d} |k|^2 u_k \bar{u}_k + \sum_{k_i \in \mathbb{Z}^d:\ \sum_{i=1}^{2q+2}(-1)^i k_i=0} u_{k_1} \bar{u}_{k_2} u_{k_3} \bar{u}_{k_4} \cdots u_{k_{2q+1}} \bar{u}_{k_{2q+2}}. \tag{3}$$

The complex symplectic form is $i \sum_k du_k \wedge d\bar{u}_k$. In order to make this construction analytic and not purely algebraic it is customary to work on spaces of analytic functions with exponential decay of the Fourier coefficients. Formally the complex Hilbert spaces, dependent of $a > 0$, $p > d/2$:

$$\bar{\ell}^{(a,p)} := \{u = \{u_k\}_{k \in \mathbb{Z}^d} \mid |u_0|^2 + \sum_{k \in \mathbb{Z}^d} |u_k|^2 e^{2a|k|}|k|^{2p} := ||u||_{a,p}^2 \le \infty\}. \tag{4}$$

Then not only the Hamiltonian H is a convergent series but most important also its associated Hamiltonian vector field is analytic, this of course is a typical situation of infinite dimensional systems coming from semi–linear PDE's, and we call such a Hamiltonian *regular analytic*.

We systematically apply the fact that we have $d + 1$ conserved quantities, generated as usual by symmetries: the d–vector *momentum* $\mathbb{M} := \sum_{k \in \mathbb{Z}^d} k|u_k|^2$ and the scalar *mass* $\mathbb{L} := \sum_{k \in \mathbb{Z}^d} |u_k|^2$ with

$$\{\mathbb{M}, u_h\} = ihu_h,\ \{\mathbb{M}, \bar{u}_h\} = -ih\bar{u}_h,\ \{\mathbb{L}, u_h\} = iu_h,\ \{\mathbb{L}, \bar{u}_h\} = -i\bar{u}_h. \tag{5}$$

The terms in equation (3) commute with $\mathbb{L}$. The conservation of momentum is expressed by the constraints $\sum_{i=1}^{2q+2}(-1)^i k_i = 0$.

1.1.2. *Choice of the tangential sites*

If in the Hamiltonian H we remove all terms of degree $2q + 2$ which do not Poisson commute with the quadratic part, we obtain a simplified Hamiltonian denoted H_{Birk} whose Hamiltonian vector field is tangent to infinitely many subspaces obtained by setting some of the coordinates equal to 0 (cf.,[13] Prop. 1). On infinitely many of them the restricted system is completely integrable, thus the first idea is to perform perturbation theory from the small solutions of these simplified systems.

The next step thus consists in choosing such a subset S which, for obvious reasons, is called of *tangential sites*.

As an example, for the cubic NLS, the condition is that we never have 3 elements of S vertices of a triangle with a right angle. This is a first condition but as one proceeds in the analysis one sees that a not wise choice of S may still create various problems, which one can in general treat as *avoidable resonances*. These resonances are described by polynomial equations in the modes S, so one avoids them by a generic choice of S. Notice though that most of the resonances are essentially created by *small frequencies* and one can give a probabilistic meaning to this statement.

With this remark in mind we partition $\mathbb{Z}^d = S \cup S^c$, $S := (j_1, \dots, j_n)$ where S are *tangential sites* and of S^c the *normal sites*. We divide $u \in \bar{\ell}^{a,p}$ in two components $u = (u_1, u_2)$, where u_1 has indexes in S and u_2 in S^c. The choice of S is subject to the constraints which make it *generic* and which are fully discussed in[13] and finally refined in[16] and completed in[15] .

We apply a standard *semi-normal form* change of variables with generating function F_{Birk} given by Formula (6) of[14] . We use the operator notation $ad(F)$ for the operator $X \mapsto \{F, X\}$. The change of variables by $\Psi^{(1)} := e^{ad(F_{Birk})}$ is well defined and analytic: $B_{\epsilon_0} \times B_{\epsilon_0} \to B_{2\epsilon_0} \times B_{2\epsilon_0}$, for ϵ_0 small enough, see[13] . This number ϵ_0 dependent of ℓ, p is the basis scale with respect to which we measure the smallness conditions.

$\Psi^{(1)}$ brings (3) to the form $H = H_{Birk} + P^4(u) + P^6(u)$ where $P^4(u)$ is of degree 4 at least cubic in u_2 while $P^6(u)$ is analytic of degree at least 6 in u, finally

$$H_{Birk} := \sum_{k \in \mathbb{Z}^d} |k|^2 u_k \bar{u}_k + \sum_{\alpha, \beta \in \mathcal{C}'} \binom{2}{\alpha} \binom{2}{\beta} u^\alpha \bar{u}^\beta, \qquad \alpha, \beta \in (\mathbb{Z}^d)^{\mathbb{N}}, \quad (6)$$

$$\mathcal{C}' : |\alpha| = |\beta| = q+1, \quad |\alpha_2| + |\beta_2| \le 2, \quad \sum_k (\alpha_k - \beta_k)k = 0,$$

$$\sum_k (\alpha_k - \beta_k)|k|^2 = 0. \quad (7)$$

The three constraints in this formula express the conservation of $\mathbb{L}, \mathbb{M}$ and of the *quadratic energy* $\mathbb{K} := \sum_{k \in \mathbb{Z}^d} |k|^2 u_k \bar{u}_k$.

At this point enters one of the main steps to start a KAM algorithm, we switch to polar coordinates, and at the same time introduce parameters for the natural quasi-periodic orbits of the simplified Hamiltonian H_{Birk}, we set for $i = 1, \dots n$:

$$u_k := z_k \text{ for } k \in S^c, \quad u_{j_i} := \sqrt{\xi_i + y_i} e^{ix_i} = \sqrt{\xi_i}(1 + \frac{y_i}{2\xi_i} + \dots) e^{ix_i} \quad (8)$$

considering the $\xi_i > 0$ as parameters $|y_i| < \xi_i$ while $y, x, w := (z, \bar{z})$ are dynamical variables.

The main point of this construction is that, when we fix the parameters $\xi \neq 0$, the conditions $z_k = 0$, $y = 0$ do not define anymore the point 0 but rather the orbit with those parameters as action variables. Thus the perturbation algorithm will be dependent on these parameters, and we hope to find a Cantor set (a fractal set) of these parameters sufficiently large where the algorithm converges producing the desired quasi–periodic solution.

The symplectic form is now $dy \wedge dx + i \sum_{k \in S^c} dz_k \wedge d\bar{z}_k$.

We give degree 0 to the angles x, degree 2 to y and 1 to w. In Taylor expansion, we develop $\sqrt{\xi_i + y_i} = \sqrt{\xi_i}(1 + \frac{y_i}{2\xi_i} + \ldots)$ as a series in $\frac{y_i}{\xi_i}$.

We still call H the composed Hamiltonian $H \circ \Psi^{(1)} \circ \Phi_\xi$. It is defined in the domain $D(s, r) \times \mathcal{O}$ where

$$D(s,r) := \{x, y, w \, : \, x \in \mathbb{T}^n_s, \, |y| \leq r^2, \, \|w\|_{a,p} \leq r\} \subset \mathbb{T}^n_s \times \mathbb{C}^n \times \ell^{(a,p)}, \quad (9)$$

$\mathbb{T}^n_s$ denotes the compact subset of the complex torus $\mathbb{T}^n_{\mathbb{C}} := \mathbb{C}^n / 2\pi \mathbb{Z}^n$ where $x \in \mathbb{C}^n$, $|\mathrm{Im}(x)| \leq s$.

The set $\mathcal{O}$ is a compact domain in the space of parameters ξ and we impose $\mathcal{O} \subseteq \varepsilon^2 [1/2, 3/2]^n$. Here $\varepsilon > 0$, $s > 0$ and $0 < r < \varepsilon/4$ are auxiliary parameters which can be chosen so that the change of variables is analytic.

Definition 1. The *normal form* $\mathcal{N}$ collects all the terms of H_{Birk} of degree ≤ 2 (dropping the constant terms). We then set $P := H - \mathcal{N}$.

In the next sections we shall discuss all the special properties of the normal form $\mathcal{N}$ and of the *perturbation* P which will allow us to perform a successful KAM algorithm whose final output will be to find a Cantor set of positive measure in the parameters ξ where the NLS Hamiltonian in suitable coordinates is quadratic, block diagonal and reduced and hence produces the desired solutions.

2. Reduction of the form

By explicit computation, and under simple genericity conditions, we have:

$$\mathcal{N} = (\omega(\xi), y) + \sum_{k \in S^c} |k|^2 |z_k|^2 + \mathcal{Q}(\xi; x, w), \quad \omega_i(\xi) := |j_i|^2 + \omega_i^{(1)}(\xi) \quad (10)$$

here $\mathcal{Q}(\xi; x, w)$ is a quadratic Hamiltonian in the variables w with coefficients trigonometric polynomials in x given by Formula (30) of[13] .

This is quite a complicated formula which still needs a lot of manipulations in order to make it amenable to our goals.

The frequency modulation $\omega^{(1)}$ is homogeneous of degree q in ξ and given by the following explicit formula. We introduce

$$A_r(\xi_1,\ldots,\xi_m) := \sum_{\sum_i k_i = r} \binom{r}{k_1,\ldots,k_m}^2 \prod_i \xi_i^{k_i} \tag{11}$$

and we have

$$\omega^{(1)}(\xi) = \nabla_\xi A_{q+1}(\xi) - (q+1)^2 A_q(\xi)\underline{1}. \tag{12}$$

In[13] we decompose this very complicated infinite dimensional quadratic Hamiltonian into infinitely many decoupled finite dimensional systems, and most important reduced to constant coefficients by an explicit symplectic change of variables which is a *combinatorial phase shift*.

This is Theorem 1 of[13] .

Theorem 1. *For all* generic *choices* $S = \{j_1,\ldots,j_n\} \in \mathbb{Z}^{nd}$ *of the tangential sites, there exists a* phase shift map $L : S^c \to \mathbb{Z}^n$, $L : k \mapsto L(k)$, $|L(k)| \leq 4qd$ *such that the analytic symplectic change of variables:*

$$\Psi : z_k = e^{-i(L(k),x)} z_k', \quad y = y' + \sum_{k\in S^c} L(k)|z_k'|^2, \quad x = x', \tag{13}$$

from $D(s,r/2) \to D(s,r)$ *has the property that* $\mathcal{N}$ *in the new variables:*

$$\mathcal{N} \circ \Psi = (\omega(\xi),y') + \sum_{k\in S^c} \tilde{\Omega}_k|z_k'|^2 + \tilde{\mathcal{Q}}(w'), \tag{14}$$

has constant coefficients, *where* $\omega(\xi)$ *is defined in* (10) *and furthermore:*

i) **Non-degeneracy** *The map* $(\xi_1,\ldots,\xi_m) \mapsto (\omega_1(\xi),\ldots,\omega_m(\xi))$ *is a diffeomorphism for* ξ *outside a homogeneous real algebraic hypersurface.*

ii) **Asymptotic of the normal frequencies:** *We have* $\tilde{\Omega}_k = |k|^2 + \sum_i |j_i|^2 L^{(i)}(k)$.

iii) **Reducibility:** *The matrix* $\tilde{Q}(\xi)$ *of the quadratic form* $\tilde{\mathcal{Q}}(\xi,w')$ *(see formula (23)) depends only on the variables* ξ, *its entries are homogeneous of degree* q *in these variables. It is block–diagonal and satisfies the following properties:*

All of the blocks except a finite number are self adjoint of dimension $\leq d+1$. *The remaining finitely many blocks have dimension* $\leq 2d+1$.

All the (infinitely many) blocks are described by a finite list of matrices $\mathcal{M}(\xi)$.

iv) **Smallness:** *If $\varepsilon^3 < r < c_1\varepsilon$, the perturbation $\tilde{P} := P \circ \Psi$ is small, more precisely we have the bounds:*

$$\|X_{\tilde{P}}\|_{s,r}^{\lambda} \leq C(\varepsilon^{2q-1}r + \varepsilon^{2q+3}r^{-1}), \tag{15}$$

where C is independent of r and depends on ε, λ only through λ/ε^2.

2.1. The matrix blocks and the geometric graph Γ_S

The phase shift L and the Hamiltonian $\tilde{Q}(\xi)$ are described in terms of a *2-coloured* marked graph Γ_S with vertices in $\mathbb{Z}^d$ and labels in $\mathbb{Z}^n$ which encodes the possible interactions between the normal frequencies $k \in S^c$.

It is convenient to describe this in a combinatorial way, see.[13]

Definition 2. We consider the free abelian group $\mathbb{Z}^n$ with canonical basis e_i and the maps $\eta : \mathbb{Z}^n \to \mathbb{Z}$, $\pi : \mathbb{Z}^n \to \mathbb{Z}^d$, $\pi^{(2)} : \mathbb{Z}^n \to \mathbb{Z}$

$$\eta : e_i \mapsto 1, \qquad \pi : e_i \mapsto j_i, \qquad \pi^{(2)} : e_i \mapsto |j_i|^2. \tag{16}$$

Definition 3 (edges). Consider the elements

$$X := \left\{ \ell := \sum_{j=1}^{2q} \pm e_{i_j} = \sum_{i=1}^{m} \ell_i e_i, \quad \ell \neq 0, -2e_i \; \forall i, \quad \eta(\ell) \in \{0, -2\} \right\}. \tag{17}$$

Notice the *mass constraint* $\sum_i \ell_i = \eta(\ell) \in \{0, -2\}$. We call all these elements respectively the *black*, $\eta(\ell) = 0$ and *red* $\eta(\ell) = -2$ *edges* and denote them by $X^{(0)}, X^{(-2)}$ respectively.

Each edge carries a *quadratic energy* (which is a positive integer):

$$K(\ell) := \frac{1+\eta(\ell)}{2}(|\pi(\ell)|^2 + \pi^{(2)}(\ell)), \quad \frac{1+\eta(\ell)}{2} = \pm\frac{1}{2}.$$

Choose now a finite set of integral vectors $S := \{j_1, \ldots, j_n\}$ in $\mathbb{Z}^d$. For the NLS these are the tangential sites but the construction is purely geometric.

Definition 4. Given $\ell \in X^{(0)}$ denote by $\mathcal{P}_\ell$ the set of pairs $h, k \in \mathbb{Z}^d$ satisfying:

$$\sum_{j=1}^{m} \ell_j j_j + h - k = 0, \quad \sum_{j=1}^{m} \ell_j |j_j|^2 + |h|^2 - |k|^2 = 0 \qquad \ell \in X^{(0)}. \tag{18}$$

If $\ell \in X^{(-2)}$ we denote by $\mathcal{P}_\ell$ the set of unordered pairs $\{h, k\}$, $h, k \in \mathbb{Z}^d$ satisfying:

$$\sum_{j=1}^{m} \ell_j j_j + h + k = 0, \quad \sum_{j=1}^{m} \ell_j |j_j|^2 + |h|^2 + |k|^2 = 0 \qquad \ell \in X^{(-2)}. \tag{19}$$

From this we build a graph. When h, k satisfy formula (18) we join them by an oriented *black* edge, marked ℓ, with source h and target $k = h + \sum_{j=1}^{m} \ell_j \mathbf{j}_j$. Formula (19) is symmetric and we join h, k by an unoriented *red* edge marked ℓ.

Note that the two conditions have a geometric meaning expressed through the quadratic energy. The first means that h lies on the hyperplane

$$H_\ell : \quad (x, \pi(\ell)) = K(\ell), \tag{20}$$

while k lies on the parallel hyperplane $H_{-\ell}$.

The second condition means that h, k are opposite points on the sphere

$$S_\ell : \quad |x|^2 + (x, \pi(\ell)) = K(\ell). \tag{21}$$

Our main object of interest are the connected components of this graph, called *geometric blocks*. In[13] we have shown that for a generic choice of S the set S is itself a connected component, called the *special component* all other components thus decompose S^c.

The phase shift is constructed by choosing in each block a special element, called *root* and then connecting each vertex k with some $L(k)$. The vector $L(k)$ tells us how to go from the root $\mathbf{r}(k)$ of the component A of the graph Γ_S to which k belongs, to k, namely:

$$k + \sum_i L_i(k) \mathbf{j}_i = \sigma(k) \mathbf{r}(k), \quad |k|^2 + \sum_i L_i(k) |\mathbf{j}_i|^2 = \sigma(k) |\mathbf{r}(k)|^2, \tag{22}$$

where $\sigma(k) = 1 + \sum_i L_i(k)$. This definition is well posed even if A is not a tree, so that one can *walk* from $\mathbf{r}(k)$ to k in several ways, from our genericity conditions.

2.1.1. *The NLS blocks*

We now use the graph introduced in order to describe $\tilde{Q}$:

$$\tilde{Q} = \sum_{k \in S^c} \omega^{(1)}(\xi) \cdot L(k) |z_k'|^2 + \sum_{\ell \in X_q^0} c(\ell) \sum_{(h,k) \in \mathcal{P}_\ell} z_h' \bar{z}_k'$$

$$+ \sum_{\ell \in X_q^{-2}} c(\ell) \sum_{\{h,k\} \in \mathcal{P}_\ell} [z_h' z_k' + \bar{z}_h' \bar{z}_k'], \tag{23}$$

where, given an edge ℓ, we set $\ell = \ell^+ - \ell^-$ and define:

$$
c_q(\ell) \equiv c(\ell) := \begin{cases} (q+1)^2 \xi^{\frac{\ell^+ + \ell^-}{2}} \displaystyle\sum_{\substack{\alpha \in \mathbb{N}^m \\ |\alpha + \ell^+|_1 = q}} \binom{q}{\ell^+ + \alpha}\binom{q}{\ell^- + \alpha} \xi_i^\alpha & \ell \in X_q^0, \\[2em] (q+1)q\xi^{\frac{\ell^+ + \ell^-}{2}} \displaystyle\sum_{\substack{\alpha \in \mathbb{N}^m \\ |\alpha + \ell^+|_1 = q-1}} \binom{q+1}{\ell^- + \alpha}\binom{q-1}{\ell^+ + \alpha} \xi_i^\alpha & \ell \in X_q^{-2}. \end{cases}
$$

(24)

We see now that the graph has been constructed in order to *decouple* $\tilde{Q} = \sum_A \tilde{Q}_A$. The sum runs over all geometric blocks $A \in \Gamma_S$ and, if $E_b(A), E_r(A)$ denotes the set of resp. black and red edges in A:

$$
\tilde{Q}_A := \sum_{k \in A} \omega^{(1)}(\xi) \cdot L(k) |z_k'|^2 + \sum_{\ell \in E_b(A)} c_q(\ell) z_h' \bar{z}_k' + \sum_{\ell \in E_r(A)} c_q(\ell) [z_h' z_k' + \bar{z}_h' \bar{z}_k'] \quad (25)
$$

is a quadratic Hamiltonian in the variables $w_A' := z_k', \bar{z}_k'$ with k running over the vertices of A, we have $\{\tilde{Q}_A, \tilde{Q}_B\} = 0, \forall A \neq B$.

We have now a list of decoupled finite dimensional quadratic Hamiltonians, which we reduce to symplectic normal form. In the cubic NLS one may choose S so that all the eigenvalues are distinct and hence the normal form is diagonal (over $\mathbb{C}$). In the general case we can prove, that by further change of variables we reduce to a block diagonal form, where distinct blocks have different eigenvalues, this is sufficient for the KAM algorithm.

3. The KAM algorithm

The starting point for our KAM Theorem is a class of *admissible* Hamiltonians H defined in $D(s,r) \times \mathcal{O}$, which we describe by axiomatizing some of the properties of our given H from which we start. It is well known that, for each $\xi \in \mathcal{O}$, the Hamiltonian equations of motion for the unperturbed $\mathcal{N}$ admit the special solutions $(x, 0, 0, 0) \to (x + \omega(\xi)t, 0, 0, 0)$ that correspond to invariant tori in the phase space.

Our aim is to prove that, under suitable hypotheses, there is a set $\mathcal{O}_\infty \subset \mathcal{O}$ of positive Lebesgue measure, so that, for all $\xi \in \mathcal{O}_\infty$ the Hamiltonians H still admit invariant tori (close to the ones of the unperturbed system). Moreover there exists a symplectic change of variables for which: the tori are defined by the conditions $y = z = 0$, the Hamiltonian vector field X_H restricted to these tori is $\sum_{i=1}^n \omega_i^\infty(\xi) \frac{\partial}{\partial x_i}$ while X_H linearized at a torus is block-diagonal in the normal variables with x-independent block matrices

$\Omega_t^\infty(\xi)$. The previous condition depends only on the terms $H^{\leq 2}$ of H of degree ≤ 2. Such tori are called *reducible KAM tori* for H.

This change of variables is obtained as the limit of a converging sequence of changes of variables, denoted KAM steps. At each step, the goal is to reduce the size of $H^{\leq 2}$, and this is done by a change of variables generated by a Hamiltonian F of degree ≤ 2, which is constructed by solving the Homological equation $\{F, \mathcal{N}\} = \tilde{H}^{\leq 2}$, where $\tilde{H}^{\leq 2}$ is the projection of $H^{\leq 2}$ on the range of $\mathrm{ad}(\mathcal{N})$.

In this procedure at each step we have to decrease both the radii r, s for different reasons, we decrease r using Cauchy type of estimates, since our operations involve derivatives through Poisson brackets, and decrease s by successive *higher frequency cuts*.

But the subtlest point is how we decrease the compact domain $\mathcal{O}$ of the parameters.

At each step of the algorithm appear algebraic singularities in the parameters ξ since we divide by functions of ξ which may vanish on some hypersurface. We thus have to *carve* some small tubular neighbourhood of each of these hypersurfaces at each step.

The game is to carve a neighbourhood large enough so to control the singularity, but small enough so to end up at the end with a set of parameters of positive measure. Since in the end we shall have carved away infinitely many regions this is a typical construction of a Cantor set, of fractal nature. The main point is to show that the final Cantor set is of positive measure, in particular non-empty.

At this point enters a subtle property of the perturbations which has to be insured for the success of the algorithm, this is the existence of the *quasi–Töplitz norm*,[18] a development of the *Töplitz–Lipschitz conditions* of[8] .

This norm, roughly speaking, estimates how far a given function of the Fourier coefficients, parametrized by $\mathbb{Z}^d$, differs from one which is *piecewise Töplitz*, i.e. piecewise translation invariant, for some finite linear stratification of the group $\mathbb{Z}^d$. The linear stratification is of combinatoric nature and becomes finer and finer as the algorithm runs. The condition allows at each step to carve away only finitely many regions which imply the remaining possibly infinite number of constraints. One should finally remark that these special analytic properties depend on the fact that our infinite dimensional Hamiltonian system if fact arises from a non linear PDE on a torus.

References

1. D. Bambusi and B. Grébert, *Duke Math. J.* **135**(3), 507 (2006).
2. M. Berti and L. Biasco, *Comm. Math. Phys.* **305**(3), 741 (2011).
3. M. Berti and Ph. Bolle, *J. European Math. Soc.* **15**(1), 229 (2013).
4. M. Berti, L. Corsi, M. Procesi, to apper in *Comm. in Mathematical Physics*.
5. M. Berti and M. Procesi, *Duke Math. J.* **159**(3), 479 (2011).
6. J. Bourgain. *Ann. of Math.* **148**(2), 363 (1998).
7. W. Craig and C.E. Wayne, *Comm. Pure Appl. Math.* **46**(11), 1409 (1993).
8. L.H. Eliasson and S.B. Kuksin, *Ann. of Math.* **172**(1), 371 (2010).
9. J. Geng, J. You, and X. Xu, *Adv. Math.* **22**(6) 5361 (2011).
10. S. Kuksin and J. Pöschel, *Ann. of Math.*, **143**(1), 149 (1996).
11. B. Van Nguyen, *Characteristic polynomials, associated to the NLS.* Ph.D. Thesis, "Sapienza" University of Rome - 2013.
12. J. Pöschel, *Ann. Scuola Norm. Sup. Pisa Cl. Sci. (4)* **23**(1), 119 (1996).
13. C. Procesi and M. Procesi, *Comm. Math. Phys.* **312**(2), 501 (2012).
14. M. Procesi and C. Procesi, *Advances in Math.* **272**, 399 (2015).
15. M. Procesi and C. Procesi, *Reducible quasi-periodic solutions for the NLS.* Preprint.
16. M. Procesi, C. Procesi, and B. Van Nguyen, *Rend. Lincei Mat. Appl.* **24**, 1 (2013).
17. W.-M. Wang, arXiv:1007.0156.
18. X. Xu and M. Procesi. *SIAM J. of Math. Anal.* **45**(4), 2148 (2013).

Chapter 9

Portfolio optimization: A mean field theory approach

G. Rotundo

Dipartimento Metodi e Modelli per l'Economia, "Sapienza" University of Rome (Rome, Italy)

B. Tirozzi

Dipartimento di Fisica, "Sapienza" University of Rome (Rome, Italy) and Enea Research Centre Frascati

This paper examines the problem of portfolio selection by the point of view of big investors that deal with a large amount N of shares and derivatives. This implies that deterministic numerical models for the portfolio optimization are inefficient because their complexity grows quickly as N increases. Thus, this paper examines a statistical mechanics approach to portfolio selection with constraints on the budget consumption and about the risk management and gives the conditions for optimal portfolio selection.

1. Introduction

The problem of portfolio selection is one of the most studied in Economics and Finance. The main model is the Markowitz's one, where a proper balance of risk and return is seeked. Since the seminal work of Markowitz, many other models have been developed, that differ for the objective function and for the constraints, that embed in the model remarks driven by the economic and financial theory. While the base model is suitable for an explicit solution -for instance through the method of the Lagrange multipliers-the management of large portfolio suffers of the curse of dimensionality. Therefore, the approach of Statistical Mechanics, that deals with systems with many units, offers perspectives and methods suitable for large dimensional porfolio problems. The aim of the problem of portfolio selection is to find the portfolio that is optimal with respect to some investment criterion. These criteria are expressed by appropriate mathematical functions in order to build the hamiltonian function of the system and the related Gibb's measure. Simulations made by the Monte Carlo Method produce a Markov chain that converges to the optimal portfolio, whose energy value has been computed apart by the estimate of the partition function by the saddle point method.

2. The formalization of portfolio optimality criteria

The Markowitz model for portfolio selection[1-3] is a classic model in Economics/Finance. Let our portfolio be described by

$$x = (x_1, \cdots, x_N) \in R^N, 0 < x_i < 1 \qquad (1)$$

and let the total available capital be equal to 1. Thus, in the general case, x_i, $i = 1, \cdots, N$ can be interpreted as the portion of the given capital that must be invested in each asset. This assumption implies that it is always possible to buy the desired amount of shares. In general, this is not obvious, because small holdings are not desiderable for several reasons: we quote the transaction costs, the minumum lot size. The assumption also implies that there are no discounts for acquiring large amounts, and this is quite a difference between the stock market and the market of goods.

We also assume that borrowing money and "short sales" is not allowed ($x_i > 0$). Let us introduce the constraints. Obviously, any portion of the capital that is not invested produces a return equal to 0, so all the available capital should be invested. This leads to the constraint

$$\sum_{i=1}^{N} x_i = 1 \qquad (2)$$

Let $r = (r_1, \cdots, r_N)$ be the vectors of returns, where r_i is the return of the asset i. Therefore, the return of the portfolio is given by $R = \sum_i r_i x_i$.

However, as soon as further constraints are introduced - for instance, constraints on the cardinality (LAM model)- the problem falls into the class of NP-hard problems. These are the reasons for looking for other approaches for the solution of the problem. Statistical mechanics deals with large numbers of units, and it is well suitable for large assets portfolios. We start our analysis from a slight modification of the classic problem.

First, we relax the constraints, considering them not strict, but represented by penalty functions. In our model, the constraint $\sum_i x_i = 1$ is substituted by:

$$f_4(x) = 0, \ f_4(x) = \frac{1}{2N}\left(1 - \sum_{i=1}^{N} x_i\right)^2 \tag{3}$$

We are going to minimize this function. Since the constraint is not strict the first order, we allow that a small amount of capital doesn't produce any return.

Let $R_N = max_{1\leq i\leq N}r_i$. We change also the constraint $\sum_i r_i x_i > \overline{R}$, substituting it by the function

$$f_2(x) = R_N - \sum_{i=1}^{N} x_i r_i \tag{4}$$

The term R_N can be substituted by any number bigger than $max_{1\leq i\leq N}r_i$. This remark is relevant because R_N could change during the time, but calculus is much simpler if we assume that R_N is constant during the period under examination.

In our setting, we change the definition of risk. Let $V = (V_1, \cdots, V_N) \in R^N$ the vector of variances of the assets, and $V = min_{1\leq i\leq N}V_i$. We assume that R_N is constant during the period under esamination. We target to the minimum of the risk

$$f_3(x) = \sum_{i=1}^{N} x_i V_i - min_{1\leq i\leq N}V_i \tag{5}$$

We add a remark on R_N and V_N. They are fixed values, but are, respectively, the maximum of the returns and the minimum of the variance, that are random variables. The extremal value theory provides the following estimate of their behavior as soon as N increases

$$V_N \simeq \sqrt{\log N}, \ R_N \simeq \sqrt{\log N} \tag{6}$$

We consider a further request on the diversification of the risk. The optimum for the function f_4 and f_2 would arise in the case in which all the capital is invested in the asset with the highest returns, but this usually implies a high risk. The request of minimum on the function f_3 does not allow this solution, but we want to introduce a function that gets its optimum when the total capital is spread all over the available assets. Thus, we introduce the following function:

$$f_1(x) = -\frac{1}{2N}\sum_{i\neq j; i,j=1}^{N} x_i x_j \tag{7}$$

that belongs to the class of concentration indices. Other possible functions are listed in.[4]

The function $f_1(x)$ gets its optimum when all the invested quantities are equal to $\frac{1}{N}$ that under the constraint $f_4(x) = 0$. This function is introduced because (under the constraint $f_4(x) = 0$) $-f_1(x)$ behaves like the entropy of the investment system discussed in the last paragraph. The investment indifferent to the assets gets the maximum of this function, while the investment concentrated just in one share leads to the minimum value. We don't consider neither the transaction costs nor the influence of the presence of options on the assets. Thus, our problem is to find the minimum of

$$\sum_{i=1}^{4} \lambda_i f_i(x), \tag{8}$$

with $\lambda_i > 0$, $i = 1, .., 4$.[a]

3. The statistical mechanics of portfolio selection

We have to deal with a large number of variables x_i, $i = 1.., N$ and there is an energy function $H_N(x, r, V)$ defined in the previous section. Here $x = (x_1, ..x_N)$ is the vector defined before $r = (r_1, ..r_N)$ the vector of the returns, $V = (V_1, \ldots, V_N)$ the vector of the variance of the assets. Since N can be very large we use the statistical mechanics approach based on the Hamiltonian:

$$H_N(x) = \frac{\lambda_1}{2N} \sum_{i \neq j} x_i x_j - \lambda_2 (R_N - \sum_i x_i r_i) +$$

$$- \lambda_3 (\sum_i x_i V_i - V_N) - \frac{\lambda_4}{2N} (1 - \sum_i x_i)^2$$

To this Hamiltonian we associate the Gibbs measure:

$$< g(x) >= \int \Pi dx_i \frac{g(x) e^{-\beta H_N(x)}}{Z_N(\beta)} \tag{9}$$

where $Z(\beta)$ is the partition function, β can be interpreted as the inverse of the temperature T, $\beta = \frac{1}{T}$, the temperature has the meaining of the measure of the internal noise of the system. Z_N has the form:

---

[a]As a first step we make also the hypothesis that the assets are independent. It is not necessary to consider any absolute value or the square of f_2 and f_3 because they are always positive. f_2 and f_3 can be seen as exact penalty functions.

$$Z_N = \int \Pi_{i=1}^N dx_i e^{-\beta H_N} \tag{10}$$

Our target is the calculus of the limit free energy

$$f = \lim_{N \to \infty} \left(-\frac{1}{\beta N} \log Z_N \right)$$

We notice that the system has two level of randomness, one given by the Gibbs measure and the other one given by the randomness of the variables r, V. Thus both the f_N and the Gibbs measure depend on these set of random variables. We have to introduce two type of averaging one already introduced before $< \dot{>}$ and the other one being the averaging with respect to the measure of the r, V variables which are assumed to be $N(0, 1)$ and $N(\mu, 1)$ respectively with $\mu > 0$. We will indicate with the symbol $<< \cdot >>$ the expectation with respect to the r, V variables

$$<< g(r, V) >> = \int_{-\infty}^{+\infty} \int_{-\infty}^{+\infty} \Pi_{i=1}^N dr_i e^{-r_i^2/2} \Pi_{i=1}^N dV_i e^{-(V_i - \mu)^2/2} g(r, V)$$

We remark that (3) allows to neglect the term $\lambda_2 R_N - \lambda_3 V_N$, because, in the limit $N \to \infty$, they contribute with the term of the order of $\frac{\sqrt{\log(N)}}{N}$, so their contribution goes to 0. We drop all the terms which are not of the order N in the hamiltonian since only terms of the order N contribute to the free-energy. Therefore, we focus on

$$H_N(x) = \frac{1}{2N}(\lambda_1 - \lambda_4)(\sum_i x_i)^2 - \sum_i h_i x_i$$

where $h_i = -\lambda_2 r_i + \lambda_3 V_i$.

In order to get the mean value of the $\sum_i x_i$ we add a linear term to the Hamiltonian

$$H_N(x) = \frac{1}{2N}(\lambda_1 - \lambda_4) \sum_i x_i x_j - \sum_i h_i x_i + u \sum_i x_i$$

Thus,

$$f_N = -\frac{1}{\beta N} \log \int_0^1 \Pi_{i=1}^N dx_i e^{-\beta H}$$

$$= -\frac{1}{\beta N} \log \int_0^1 \Pi_{i=1}^N dx_i e^{\frac{\beta}{2N}(\lambda_4 - \lambda_1) \sum_i x_i x_j + \beta \sum_i (h_i - u) x_i}$$

The utility of the free-energy is that, when it is analytic, i.e. if there are no phase transitions, it is possible to get the order parameters of the theory by derivation. So we have

$$< m >= \frac{\partial f}{\partial u}|_{u=0} = \lim_{N\to\infty} \frac{1}{N} \sum_i < x_i >$$

$$< f_2 >= \frac{\partial f}{\partial \lambda_2}|_{u=0} = \lim_{N\to\infty} \frac{1}{N} \sum_i r_i < x_i >$$

$$< f_3 >= -\frac{\partial f}{\partial \lambda_3}|_{u=0} = \lim_{N\to\infty} \frac{1}{N} \sum_i V_i < x_i >$$

$$< f_4 >= -\frac{\partial f}{\partial \lambda_4}|_{u=0} = \lim_{N\to\infty} \frac{1}{N} < \frac{1}{2N}(\sum_i x_i)^2 >$$

$$< f_1 >= - < f_4 > .$$

It is evident from this formula that all the order parameters are random variables depending on the r, V variables in a complicated way. But in the next section we are going to show that the free-energy is self-averaging, i.e. that in the limit $N \to \infty$ it does not depend on the values of the r, V variables. With a simple argument one can show that also the derivatives with respect to the parameters of the theory are self-averaging and so we get that also that order parameters are not depending, in probability, on the choice of the r and V variables.

Now we show how to compute the partition function of Z_N in order to get the limit free-energy. Define $\lambda = \lambda_4 - \lambda_1 > 0$ we have:

$$Z_N = \int \Pi_{i=1}^N dx_i e^{-\beta H_N} = \int \Pi_{i=1}^N dx_i e^{\frac{\beta}{2N}\lambda(\sum_i x_i)^2 + \beta \sum_i (h_i - u)x_i}.$$

Remembering that

$$e^{\frac{\beta\lambda}{2N}(\sum_i x_i)^2} = \int \frac{dz}{\sqrt{2\pi}} e^{-\frac{z^2}{2} + z\sqrt{\frac{\beta\lambda}{N}} \sum_i x_i},$$

we get

$$Z_N = \int \Pi_{i=1}^N dx_i e^{-\beta H_N} = \int \frac{dz}{\sqrt{2\pi}} e^{-\frac{z^2}{2}} \Pi_{i=1}^N \int dx_i e^{(z\sqrt{\frac{\beta\lambda}{N}} + \beta(h_i - u))x_i}.$$

It is evident from the expression of the Hamiltonian and from this expression of the partition function Z_N that we are dealing with a statistical mechanics mean field system with a particular Gaussian random field. As in that case we change variable $z = \sqrt{\beta N \lambda} v$. We expand the exponential and the logarithm in the expressions below using the hypothesis of β small enough, i.e. we study the system in the high temperature regime.

$$Z_N = \sqrt{\beta N \lambda} \int \frac{dv}{\sqrt{2\pi}} e^{-\frac{\beta N \lambda}{2} v^2 + \sum_i \log \int_0^1 dx_i e^{\beta(\lambda v + h_i - u)x_i}}$$

$$= \sqrt{\beta N \lambda} \int \frac{dv}{\sqrt{2\pi}} e^{-\frac{\beta N \lambda}{2} v^2 + \sum_i \log \int_0^1 dx_i (1 + \beta(\lambda v + h_i - u)x_i)}$$

$$= \sqrt{\beta N \lambda} \int \frac{dv}{\sqrt{2\pi}} e^{-\frac{\beta N \lambda}{2} v^2 + \sum_i \log(1 + \frac{1}{2}\beta(\lambda v + h_i - u))}$$

For small β we can use the approximation $\log(1 + x) = x - \frac{x^2}{2}$

$$= \sqrt{\beta N \lambda} \int \frac{dv}{\sqrt{2\pi}} e^{-\frac{\beta N \lambda}{2} v^2 + \sum_i (\frac{1}{2}\beta(\lambda v + h_i - u) - \frac{1}{8}\beta^2(\lambda v + h_i - u)^2)}$$

Let us define

$$G(h_i) = \frac{1}{2}\beta(\lambda v + h_i - u) - \frac{1}{8}\beta^2(\lambda v + h_i - u)^2,$$

we can write

$$Z_N = \sqrt{\beta N \lambda} \int \frac{dv}{\sqrt{2\pi}} e^{-\frac{\beta N \lambda}{2} v^2 + \sum_i G(h_i)}.$$

Assuming the strong-self averaging property ,see the remark at the end of this section, in the limit $N \to \infty$, $\sum_{i=1}^N G(h_i) \to N << G(h_i) >>$, then we have

$$Z_N = \sqrt{\beta N \lambda} \int \frac{dz}{\sqrt{2\pi}} e^{-\frac{\beta N \lambda}{2} v^2 + N << G(h_i) >>}.$$

We introduce the constants $c_1 = << h_i >>$ and $c_2 = << h_i^2 >> = c_2$, so that $c_1 = \lambda_3 \mu$ and $c_2 = \lambda_2^2 + \lambda_3^2(1 + \mu^2)$. Then

$$<< G(h_i) >> = \frac{\beta}{2}(\lambda v + c_1 - u) - \frac{\beta^2}{8}(\lambda^2 v^2 + c_2 + u^2 + 2\lambda v c_1 - 2\lambda v u - 2c_1 u).$$

Therefore, the partition function becomes:

$$Z_N = \sqrt{\beta N \lambda} e^{\frac{N}{2}\beta\left[(c_1-u)-\frac{\beta}{4}(c_2+u^2-2c_1u)\right]} \int \frac{dv}{\sqrt{2\pi}} e^{-\frac{N}{2}\beta\left(v^2\lambda(1+\frac{\beta}{4}\lambda)-v\lambda(1-\frac{\beta}{2}c_1+\frac{\beta}{2}u)\right)}.$$

We need only to compute the Gaussian integral on the v variable

$$Z_N = \sqrt{\beta N \lambda} \sqrt{\frac{1}{2\frac{N\beta}{2}\lambda(1+\frac{\beta}{4}\lambda)}} e^{\frac{N}{2}\beta\left[(c_1-u)-\frac{\beta}{4}(c_2+u^2-2c_1u)\right]+\frac{N}{2}\beta\lambda\left(\frac{1-\frac{\beta}{2}(c_1-u)}{2\sqrt{(1+\frac{\beta}{4}\lambda)}}\right)^2}$$

Substituting $c_1 = \lambda_3\mu$ and $c_2 = \lambda_2^2 + \lambda_3^2(1 + \mu^2)$ the limit free energy has the expression:

$$f = -\frac{1}{2}\left[(\lambda_3\mu - u) - \frac{\beta}{4}(\lambda_2^2 + \lambda_3^2(1 + \mu^2) + u^2 - 2\lambda_3\mu u)\right]$$
$$-\frac{1}{8}\lambda\left(\frac{1-\frac{\beta}{2}(\lambda_3\mu - u)}{\sqrt{(1+\frac{\beta}{4}\lambda)}}\right)^2.$$

We can compute the order parameter m

$$m = \frac{\partial f}{\partial u}\Big|_{u=0} = \frac{1}{2} - \frac{\beta}{4}\lambda_3\mu - \frac{1}{8}\frac{\lambda}{1+\frac{\beta}{4}\lambda}\beta\left(1 - \frac{\beta}{2}\lambda_3\mu\right)$$

We look for phase transitions $m = 0$. We get a linear equation

$$\beta_c = \frac{2}{\lambda_3\mu}$$

Thus we have a critical temperature at $\beta = \beta_c$ if $\lambda_3\mu > 0$, in particular if the expectation value of the V_i is positive.

Remark 1 In practice we have used the strong averaging property:

$$<< \log Z_N >> = \log << Z_N >>$$

which holds for low values of β. We didn't prove this statement but in the next section we show the self-averaging of the free-energy which is a weaker result but as important as this property.

Remark 2 The existence of a critical temperature β_c depends on the particular condition of the model parameters. It would be interesting to investigate the economical meaning of the condition. In analogy with the computation of the mean investment rate we can compute the average values of the terms appearing in the hamiltonian by deriving the expression of the free-energy with respect to $\lambda_1, \lambda_2, \lambda_3, \lambda_4$. Probably we will get other phase transitions with some economical interest. But this will be the object of further investigation.

4. Self-averaging

In the previous calculations we have made use of the self-averaging of the free-energy an important property which is formulated as follows

$$\lim_{N\to\infty} <<(f_N- <<f_N>>)^2>> \to 0$$

This property is very important and it has been proven in the theory of spin-glass and Hopfield models. It has been also shown that from the self-averaging of the free energy it follows the self-averaging of the order parameters, like m in our case. We follow the proof given in the paper.[7] Let $\mathcal{R}_k^N$ the σ algebra generated by the random variables $r_i, V_i, i = k, .., N$ and let F_N^k be the conditional expectation of $F_N = -\frac{1}{\beta}\log Z_N$ with respect to $\mathcal{R}_k^N$

$$F_N^k = E(F_N|\mathcal{R}_k^N) \equiv E_{<k}(F_N)$$

In other words

$$E_{<k}(F_N) \equiv \int \Pi_{i=1}^{k-1}p(dr_i)p(dV_i)F_N$$

From this definition it follows that F_N^k , $k = 1, .., N$ has the martingale property

$$E(F_N|\mathcal{R}_k^N) = \begin{cases} F_N^k \text{ if } k > l; \\ F_N^l \text{ if } k \le l. \end{cases}$$

Let us define

$$\Psi^k = F_N^k - F_N^{k+1}$$

So we have $F_N^1 = -\frac{1}{\beta}\log Z_N$, $F_N^{N+1} = E(-\frac{1}{\beta}\log Z_N)$, we have

$$f_N - Ef_N = \frac{1}{N}\sum_{k=1}^{N}\Psi^k$$

Applying the martingale property one gets

$$E(f_N - Ef_N)^2 = \frac{1}{N^2} \sum_{k=1}^{N} E(\Psi^k)^2$$

So, in order to prove our statement it is enough to show that:

$$E(\Psi^k)^2 \le C$$

At this aim let us define the following Hamiltonians:

$$H_k = H = -\frac{1}{2N}(\lambda_1 - \lambda_4) \sum_i x_i x_j + \sum_{i \ne k} h_i x_i$$

$$R_k(t) = t h_k x_k$$

$$\Phi(t)_k = H_k + t R_k$$

So $\Phi(t)$ is an interpolating hamiltonian such that $\Phi(1) = H$. Let us define also

$$\tilde{f}_k(t) = -\frac{1}{\beta}(\log Z_N(\Phi_k(t)) - \log Z_N(\Phi(0)))$$

From these defintions it follows that

$$\Psi^k = E_{<k}\tilde{f}_k(1) - E_{<k+1}\tilde{f}_k(1)$$

$\tilde{f}_k(0) = 0$ and also the inequality $E(\Psi^k)^2 \le 2E(\tilde{f}_k(1))^2$. From the convexity of the free-energy it follows that

$$\frac{d^2}{dt^2}(\tilde{f}_k(t)) \le 0.$$

From this inequality it follows that

$$\tilde{f}_k(1) \le \tilde{f}_k(0)' = \frac{1}{N} < x_k >_{\Phi(0)} .$$

Finally

$$E((\tilde{f}_k(1))^2 \le E(\frac{1}{N} < x_k >_{\Phi(0)})^2 < C$$

and so the theorem is proven.

5. Conclusion

We have chosen a mean-field model of statistical mechanics for investigating the behaviour of the market when the number of investors is asymptotically large in such a way that we can apply the methods of statistical mechanics. The model has two random fields which and we have proven the self-averaging property of the free-energy with respect to this randomness. But we had to use the strong self-averaging property in order to solve completely the model. There are also conditions on the model parameters in order that m should be positive and less than 1. Thus we get much more conditions in this theory than in the ususal statistical mechanics models. The analytical solution of the model can be interesting for economic applications since we can get also the typical values of the 4 constraints. We are going to analyze all these consequences in a future work.

References

1. H. M. Markowitz, *J. Finance* **7**(1):77 doi:10.2307/2975974, JSTOR 2975974 (1952).
2. H. M. Markowitz, *Portfolio Selection: Efficient diversification of investments. Cowles Foundation for Research in Economics at Yale University*, Monograph 16, John Wiley & Sons Inc., New York, (1959).
3. H. M. Markowitz, *Mean-variance analysis in portfolio choice and capital markets*, Basil Blackwell, Oxford (1987).
4. Zbigniew Michalewicz, *Genetic Algorithms + Data Structures = Evolution Programs.* Springer (1992).
5. M. Raffa, G. Rotundo, B. Tirozzi, Quaderni del Dipartimento di Matematica per le Decisioni Economiche, Finanziarie ed Assicurative, quaderno n.9 (2000).
6. M. Raffa, G. Rotundo, B. Tirozzi, *Europhysics conference abstracts of the conference Applications of Physics in Financial Analysis*, Liege, Belgiumn, July 13th-15th, (2000).
7. M. Shcherbina, B. Tirozzi, *J. Stat.Phys* **72**, 113 (1993).

Chapter 10

Fluctuations of the eigenvalue number in the fixed interval for β-models with $\beta = 1, 2, 4$

M. Shcherbina

Institute for Low Temperature Physics (Kharkov, Ukraine)

We study the fluctuation of the eigenvalue number of any fixed interval $\Delta = [a, b]$ inside the spectrum for β- ensembles of random matrices in the case $\beta = 1, 2, 4$. We assume that the potential V is polynomial and consider the cases of any multi-cut support of the equilibrium measure. It is shown that fluctuations become Gaussian in the limit $n \to \infty$, if they are normalized by $\pi^{-2} \log n$.

1. Introduction and main results

Consider β-ensemble of random matrices, whose joint eigenvalue distribution is

$$p_{n,\beta}(\lambda_1, ..., \lambda_n) = Q_{n,\beta}^{-1}[V] \prod_{i=1}^{n} e^{-n\beta V(\lambda_i)/2} \prod_{1 \le i < j \le n} |\lambda_i - \lambda_j|^\beta \qquad (1)$$

where $Q_{n,\beta}[V]$ is a normalizing factor

$$Q_{n,\beta}[V] = \int \prod_{i=1}^{n} e^{-n\beta V(\lambda_i)/2} \prod_{1 \le i < j \le n} |\lambda_i - \lambda_j|^\beta d\bar{\lambda}.$$

The function V (called the potential) is a real valued Hölder function satisfying the condition

$$V(\lambda) \ge 2(1 + \epsilon) \log(1 + |\lambda|). \qquad (2)$$

Below we denote

$$E\{(\dots)\} = \int (\dots) p_{n,\beta}(\lambda_1, ..., \lambda_n) d\bar{\lambda}. \qquad (3)$$

This distribution can be considered for any $\beta > 0$, but the cases $\beta = 1, 2, 4$ are especially important, since they correspond to real symmetric, hermitian, and symplectic matrix models respectively.

It is known (see[1,7]) that if V' is a Hölder function, then the empirical spectral distribution

$$n^{-1} \sum_{j=1}^{n} \delta(\lambda - \lambda_i)$$

converges weakly in probability defined by (1) to the function ρ (equilibrium density) with a compact support σ. The density ρ maximizes the functional, defined on the class $\mathcal{M}_1$ of positive unit measures on $\mathbb{R}$

$$\mathcal{E}_V(\rho) = \max_{m \in \mathcal{M}_1} \left\{ \int \log|\lambda - \mu| dm(\lambda) dm(\mu) - \int V(\lambda) m(d\lambda) \right\} = \mathcal{E}[V]. \quad (4)$$

The support σ and the density ρ are uniquely defined by the conditions:

$$v(\lambda) := 2 \int \log|\mu - \lambda| \rho(\mu) d\mu - V(\lambda) = \sup v(\lambda) := v^*, \quad \lambda \in \sigma, \tag{5}$$
$$v(\lambda) \leq \sup v(\lambda), \quad \lambda \notin \sigma, \qquad\qquad \sigma = \text{supp}\{\rho\}.$$

We are interested in the behavior of linear eigenvalue statistics, i.e.

$$\mathcal{N}_n[h] = \sum_{j=1}^{n} h(\lambda_j^{(n)}), \tag{6}$$

In the case of smooth test function h the behavior of $\mathcal{N}_n[h]$ now is very well understood for any $\beta > 0$. It was proven in[7] that for one cut (i.e., $\sigma = [a, b]$) polynomial potentials of generic behavior and sufficiently smooth h (8 derivatives), if we consider the characteristic functional of $\mathcal{N}_n[h]$ in the form

$$\Phi_{n,\beta}[x, h] = E\left\{ e^{x(\mathcal{N}_n[h] - E\{\mathcal{N}_n[h]\})} \right\}, \tag{7}$$

then

$$\lim_{n \to \infty} \Phi_{n,\beta}[h] = \exp\left\{ \frac{x^2}{2\beta} (\overline{D}_\sigma h, h) \right\},$$

where the "variance operator" $\overline{D}_\sigma$ and the measure ν have the form

$$(\overline{D}_\sigma h, h) = \int_\sigma \frac{h(\lambda) d\lambda}{\pi^2 X_\sigma^{1/2}(\lambda)} \int_\sigma \frac{h'(\mu) X_\sigma^{1/2}(\mu) d\mu}{\lambda - \mu},$$
$$X_\sigma(\lambda) = (b - \lambda)(\lambda - a).$$

The method of[7] was improved in[10], where it was generalized to the case of non polynomial real analytic potentials V and the test functions with 4 derivatives, and then improved once more in[16], where the case of non analytic V was also studied. The case of multi-cut (i.e. σ consisting of more than one interval) real analytic potentials was studied in[17] , where it was shown that in this case fluctuations become non Gaussian.

But the method, used in the case of smooth h, does not work in the case of h which have jumps. In particular, the method is not applicable to $h = 1_\Delta$, $\Delta = [a, b] \subset \sigma$, which means that $\mathcal{N}_n[h]$ is a number of eigenvalues inside the interval Δ. Moreover, it is known that for gaussian unitary and gaussian orthogonal ensembles (GUE and GOE) the variance of the eigenvalue number is proportional to $\log n$, while in the case of smooth test functions the variance is $O(1)$. Thus, it is hard to believe that the central limit theorem (CLT) for indicators can be obtained by methods similar to that for smooth test functions.

Till now there are only few results on the CLT for indicators. The case of GUE was studied a long time ago (see,e.g.,[8]). In the paper[3] it was shown that the Gaussian fluctuations for GUE imply similar results for GOE and GSE (i.e., the cases when $V(\lambda) = \lambda^2/2$ and $\beta = 1$ and $\beta = 4$). Even for classical random matrix models, like the Wigner model with non gaussian entries, CLT for functions with jumps was proven (see[2]) only for the Hermitian case ($\beta = 2$), and only under the assumption that the first four moments of the entries coincide with that of GUE. There are also a number of publications where CLT for the determinantal point processes are proven (see[18] and references therein or[6]). Similar results for some special kind of Pfaffian point processes were obtained in.[9]

At the present paper we use the representation of the characteristic functional of $\mathcal{N}_n[h]$ in the form of the Fredholm determinant of some operator in order to prove CLT for the indicator test functions in the case of β- models with $\beta = 1, 2, 4$. Unfortunately, since similar representations are not known for general β, the method does not work for $\beta \neq 1, 2, 4$.

Let us start form the case $\beta = 2$.

Given potential V, introduce the weight function $w_n(\lambda) = e^{-nV(\lambda)}$, and consider polynomials orthogonal on $\mathbb{R}$ with the weight w_n i.e.,

$$\int p_l^{(n)}(\lambda)p_m^{(n)}(\lambda)w_n(\lambda)d\lambda = \delta_{l,m}. \tag{8}$$

It will be used below also that $\{p_l^{(n)}\}_{l=0}^n$ satisfy the recursion relation

$$\lambda p_l^{(n)}(\lambda) = a_{l+1}p_{l+1}^{(n)}(\lambda) + b_n p_l^{(n)}(\lambda) + a_l p_l^{(n)}(\lambda). \tag{9}$$

Then consider the orthonormalized system

$$\psi_l(\lambda) = e^{-nV(\lambda)/2}p_l^{(n)}(\lambda), \ l = 0, ..., \tag{10}$$

and construct the function

$$K_n(\lambda, \mu) = \sum_{l=0}^{n-1} \psi_l(\lambda)\psi_l(\mu). \tag{11}$$

This function is known as a reproducing kernel of the system (10). It is known (see, e.g.,[8]) that for any x and any bounded integrable test functions h the characteristic functional defined by (7) for $\beta = 2$ takes the form

$$\Phi_{n,2}[x, h] = e^{-xE\{\mathcal{N}_n[h]\}} \det\Big\{1 + (e^{xh} - 1)K_n\Big\},$$

where the operator $(e^{xh} - 1)K_n$ has the kernel

$$((e^{xh} - 1)K_n)(\lambda, \mu) := (e^{xh(\lambda)} - 1)K_n(\lambda, \mu)$$

In particular, if $h = 1_\Delta$ and we set $x_n = x\pi/\log^{1/2} n$, then $\Phi_{n,2}[x_n, 1_\Delta]$ takes the form

$$\hat{\Phi}_{n,2}(x) := e^{-x_n E\{\mathcal{N}_n[1_\Delta]\}} E\{e^{x_n \mathcal{N}_n[1_\Delta]}\}$$

$$= e^{-x_n E\{\mathcal{N}_n[1_\Delta]\}} \det\Big\{1 + (e^{x_n} - 1)K_n[\Delta]\Big\}, \tag{12}$$

where

$$K_n[\Delta](\lambda, \mu) := 1_\Delta(\lambda)K_n(\lambda, \mu)1_\Delta(\mu). \tag{13}$$

Representation (12) allows us to prove CLT for the indicator test function in the case $\beta = 2$ (see, e.g.,[13]):

Theorem 1. *Let the matrix model be defined by (1) with $\beta = 2$ and real analytic potential $V(\lambda) >> \log|\lambda^2 + 1|$. Let also $\Delta = [a,b] \subset \sigma^\circ$ (here and below σ° means the internal part of the support σ of the equilibrium measure) and $x_n = x\pi \log^{-1/2} n$. Then*

$$\lim_{n\to\infty} \log \hat{\Phi}_{n,2}(x) = x^2/2. \tag{14}$$

Although the result is not new, its proof is an important ingredient of the proofs of CLT for the cases $\beta = 1, 4$, hence the proof is given in the beginning of Section 2.

For $\beta = 1, 4$ the situation is more complicated. It was shown in[19] that the characteristic functionals $\hat{\Phi}_{n,1}(x)$ and $\hat{\Phi}_{n,4}(x)$ can be expressed in terms of some matrix kernels (see (16) – (21) below). But the representation is less convenient than (8) – (12). It makes difficult the problems, which for $\beta = 2$ are just simple exercises.

We have

$$\hat{\Phi}_{n,1}(x) = e^{-x_n E\{\mathcal{N}_n[1_\Delta]\}} \det^{1/2}\left\{1 + (e^{x_n} - 1)K_{n,1}[\Delta]\right\}, \qquad (15)$$

$$\hat{\Phi}_{n,4}(x) = e^{-x_n E\{\mathcal{N}_{n/2}[1_\Delta]\}} \det^{1/2}\left\{1 + (e^{x_n} - 1)K_{n,4}[\Delta]\right\},$$

where similarly to the case $\beta = 2$ the operators $K_{n,1}[\Delta]$ and $K_{n,4}[\Delta]$ are the projection on the interval Δ of some matrix operators $K_{n,1}$, $K_{n,4}$:

$$K_{n,1}[\Delta](\lambda, \mu) = 1_\Delta(\lambda)K_{n,1}(\lambda, \mu)1_\Delta(\mu),$$

$$K_{n,4}[\Delta](\lambda, \mu) = 1_\Delta(\lambda)K_{n,4}(\lambda, \mu)1_\Delta(\mu).$$

The matrix operators $K_{n,1}$ and $K_{n,4}$ have the form

$$K_{n,1} := \begin{pmatrix} \mathcal{S}_{n,1} & \mathcal{D}_{n,1} \\ \mathcal{I}_{n,1} - \epsilon & \mathcal{S}_{n,1}^T \end{pmatrix}, \qquad \beta = 1, \ n - \text{even}, \qquad (16)$$

$$K_{n,4} := \frac{1}{2}\begin{pmatrix} \mathcal{S}_{n,4} & \mathcal{D}_{n,4} \\ \mathcal{I}_{n,4} & \mathcal{S}_{n,4}^T \end{pmatrix}, \qquad \beta = 4, \qquad (17)$$

where the entries are integral operators in $L_2[\mathbb{R}]$ with the kernels

$$\mathcal{S}_{n,1}(\lambda, \mu) = -\sum_{j,k=0}^{n-1} \psi_j(\lambda)(M_n^{(n)})_{jk}^{-1}(\epsilon\psi_k)(\mu), \quad \mathcal{S}_{n,1}^T(\lambda, \mu) = \mathcal{S}_{n,1}(\lambda, \mu),$$

$$(18)$$

$$\mathcal{D}_{n,1}(\lambda, \mu) = -\frac{\partial}{\partial\mu}\mathcal{S}_{n,1}(\lambda, \mu), \quad \mathcal{I}_{n,1}(\lambda, \mu) = (\epsilon\mathcal{S}_{n,1})(\lambda, \mu),$$

$$\mathcal{S}_{n/2,4}(\lambda, \mu) = -\sum_{j,k=0}^{n-1} \psi_j'(\lambda)(D_n^{(n)})_{jk}^{-1}\psi_k^{(n)}(\mu), \quad \mathcal{S}_{n/2,4}^T(\lambda, \mu) = \mathcal{S}_{n/2,4}(\lambda, \mu),$$

$$(19)$$

$$\mathcal{D}_{n,4}(\lambda, \mu) = -\frac{\partial}{\partial\mu}\mathcal{S}_{n,4}(\lambda, \mu), \quad \mathcal{I}_{n,4}(\lambda, \mu) = (\epsilon\mathcal{S}_{n,4})(\lambda, \mu),$$

$$\epsilon(\lambda - \mu) = \frac{1}{2}\text{sgn}(\lambda - \mu). \qquad (20)$$

Here the function $\{\psi_j\}_{j=0}^n$ are defined by (10), sgn denotes the standard signum function, and $D_n^{(n)}$ and $M_n^{(n)}$ in (18) and (19) are the left top corner $n \times n$ blocks of the semi-infinite matrices that correspond to the differentiation operator and to some integration operator respectively.

$$D_\infty^{(n)} := (\psi_j', \psi_k)_{j,k\geq 0}, \quad D_n^{(n)} = \{D_{jk}^{(n)}\}_{j,k=0}^{n-1}, \qquad (21)$$

$$M_\infty^{(n)} := (\epsilon\psi_j, \psi_k)_{j,k\geq 0}, \quad M_n^{(n)} = \{M_{jk}^{(n)}\}_{j,k=0}^{n-1}.$$

Remark 1. From the structure of the kernels it is easy to see the cases $\beta = 1, 4$ the characteristic functional can be written in the form

$$\hat{\Phi}_{n,4}(x) = \det{}^{1/2}\Big\{ J + (e^{x_n} - 1)\hat{A}_{n,1}[\Delta] \Big\} e^{-x_n E\{\mathcal{N}_n[1_{\Delta_a}]\}},$$

$$\hat{\Phi}_{n,4}(x) = \det{}^{1/2}\Big\{ J + (e^{x_n} - 1)\hat{A}_{n,4}[\Delta] \Big\} e^{-x_n E\{\mathcal{N}_n[1_{\Delta_a}]\}},$$

where $\hat{A}_{n,1} = \mathcal{S}_{n,1}J$, $\hat{A}_{n,4} = \mathcal{S}_{n,4}J$ are skew symmetric matrices

$$(A_{n,1}[\Delta])^* = -A_{n,1}[\Delta], \quad (A_{n,4}[\Delta])^* = -A_{n,4}[\Delta], \quad J = \begin{pmatrix} 0 & I \\ -I & 0 \end{pmatrix}.$$

The main problem of studying of $\hat{\Phi}_{n,1}(x)$ and $\hat{\Phi}_{n,4}(x)$ is that the corresponding operators $K_{n,1}$ and $K_{n,4}$ (differently from the case $\beta = 2$) are not self adjoint, thus even if we know the location of eigenvalues of $K_{n,1}$ and $K_{n,4}$ we cannot say something about the location of eigenvalues of $K_{n,1}[\Delta]$ and $K_{n,4}[\Delta]$.

The idea is to prove that the eigenvalue problems for $K_{n,1}[\Delta]$ and $K_{n,4}[\Delta]$ can be reduced to the eigenvalue problem for $K_n[\Delta]$ with some finite rank perturbation. For this aim we use the result of,[20] where it was observed that if V is a rational function, in particular, a polynomial of degree $2m$, then the kernels $\mathcal{S}_{n,1}, \mathcal{S}_{n,4}$ can be written as

$$\mathcal{S}_{n,1}(\lambda, \mu) = K_n(\lambda, \mu) + n \sum_{j,k=-(2m-1)}^{2m-1} F_{jk}^{(1)} \psi_{n+j}(\lambda) \epsilon \psi_{n+k}(\mu), \qquad (22)$$

$$\mathcal{S}_{n/2,4}(\lambda, \mu) = K_n(\lambda, \mu) + n \sum_{j,k=-(2m-1)}^{2m-1} F_{jk}^{(4)} \psi_{n+j}(\lambda) \epsilon \psi_{n+k}(\mu),$$

where $F_{jk}^{(1)}$, $F_{jk}^{(4)}$ can be expressed in terms of the matrix T_n^{-1}, where T_n is the $(2m-1) \times (2m-1)$ block in the bottom right corner of $D_n^{(n)} M_n^{(n)}$, i.e.,

$$(T_n)_{jk} := (D_n^{(n)} M_n^{(n)})_{n-2m+j,n-2m+k}, \qquad 1 \le j, k \le 2m - 1. \qquad (23)$$

The representation was used before to study local regimes for real symmetric and symplectic matrix models. The main technical obstacle there was the problem to prove that $(T_n^{-1})_{jk}$ are bounded uniformly in n. The problem was solved initially for the case of monomial $V(\lambda) = \lambda^{2m}$ in,[5] then for general one-cut real analytic V in[14] and finally for the general multi cut potential in,[15] where it was shown that for generic real analytic potential

V

$$|F_{jk}^{(1)}| \leq C, \quad |F_{jk}^{(4)}| \leq C. \tag{24}$$

To formulate the main results, let us state our conditions.

C1. V is a polynomial of degree $2m$ with a positive leading coefficient, and the support of its equilibrium measure is

$$\sigma = \bigcup_{\alpha=1}^{q} \sigma_\alpha, \quad \sigma_\alpha = [E_{2\alpha-1}, E_{2\alpha}] \tag{25}$$

C2. The equilibrium density ρ can be represented in the form

$$\rho(\lambda) = \frac{1}{2\pi} P(\lambda) \Im X^{1/2}(\lambda + i0), \quad \inf_{\lambda \in \sigma} |P(\lambda)| > 0, \tag{26}$$

where

$$X(z) = \prod_{\alpha=1}^{2q} (z - E_\alpha), \tag{27}$$

and we choose a branch of $X^{1/2}(z)$ such that $X^{1/2}(z) \sim z^q$, as $z \to +\infty$. Moreover, the function v defined by (5) attains its maximum only if λ belongs to σ.

Remark 2. It is known (see, e.g., [13, Theorem 11.2.4]) that for any analytic V the equilibrium density ρ always has the form (26) – (27). The function P in (26) is analytic and can be represented in the form

$$P(z) = \frac{1}{2\pi i} \oint_L \frac{V'(z) - V'(\zeta)}{(z - \zeta) X^{1/2}(\zeta)} d\zeta.$$

Hence condition C2 means that ρ has no zeros in the internal points of σ and behaves like square root near the edge points. This behavior of ρ is usually called generic.

Theorem 2. *Consider the matrix model (1) with $\beta = 1$ and even n and V satisfying conditions C1,C2. Let the interval $\Delta = [a, b] \subset \sigma^\circ$, and let the characteristic functional $\hat{\Phi}_{n,1}(x)$ be defined by (12) for $\beta = 1$ with $x_n = x\pi \log^{-1/2} n$. Then*

$$\lim_{n \to \infty} \log \hat{\Phi}_{n,1}(x) = x^2.$$

Theorem 3. *Consider the matrix model (1) with $\beta = 4$ and V satisfying conditions C1,C2. Let the interval $\Delta = [a,b] \subset \sigma^\circ$ and let characteristic functional $\hat{\Phi}_{n,4}(x)$ be defined by (12) for $\beta = 4$ with $x_n = x\pi \log^{-1/2} n$. Then*

$$\lim_{n\to\infty} \log \hat{\Phi}_{n,4}(x) = x^2/4.$$

2. Proofs

Proof of Theorem 1 Set $F(x_n) := \log \hat{\Phi}_{n,2}(x)$ and consider the Taylor expansion of $F(x_n)$ with respect to x_n up to the second order

$$F(x_n) = \frac{x_n^2}{2} \text{Tr } K_n[\Delta](1 - K_n[\Delta]) + \frac{x_n^3}{6} \text{Tr } K_n[\Delta](1 - K_n[\Delta])\tilde{R}(K_n[\Delta]),$$
$$(28)$$

$$C_1 \le \tilde{R}(t) \le C_2, \quad t \in [0,1].$$

Lemma 1.

$$\text{Tr } K_n[\Delta](1 - K_n[\Delta]) = \int_\Delta d\lambda \int_{\bar{\Delta}} K_n^2(\lambda, \mu) d\mu = \pi^{-2} \log n(1 + o(1)). \quad (29)$$

The lemma implies that the first term in the r.h.s. of (28) tends to $x^2/2$, while the second one is bounded by $cx_n^3 \log n = o(1)$, since

$$C_1 \text{Tr } K_n[\Delta](1 - K_n[\Delta]) \le \text{Tr } K_n[\Delta](1 - K_n[\Delta])\tilde{R}(K_n[\Delta])$$
$$\le C_2 \text{Tr } K_n[\Delta](1 - K_n[\Delta]). \quad (30)$$

Hence we get the Theorem 1 assertion. Thus we are left to prove Lemma 1.

Proof of Lemma 1. Take $d_n = \log^{1/3} n$ and write

$$\text{Tr } K_n[\Delta](1 - K_n[\Delta]) = \int_\Delta d\lambda \int_{\bar{\Delta}} d\mu K_n^2(\lambda, \mu)$$

$$= \left(\int_a^{a+d_n} d\lambda + \int_{a+d_n}^{b-d_n} d\lambda + \int_{b-d_n}^b d\lambda \right) \quad (31)$$

$$\times \left(\int_{a-d_n}^a d\mu + \int_b^{b+d_n} d\mu + \int_{-\infty}^{a-d_n} d\mu + \int_{b+d_n}^\infty d\mu \right) K_n^2(\lambda, \mu).$$

The Christoffel-Darboux formula implies

$$\int K_n^2(\lambda, \mu)(\lambda - \mu)^2 d\lambda d\mu = a_n \int (\psi_n(\lambda)\psi_{n-1}(\mu) - \psi_n(\mu)\psi_{n-1}(\lambda))^2 d\lambda d\mu$$

$$= 2a_n \le C, \quad (32)$$

where a_n is the recursion coefficient of (9), and we have used the result of[11] (see also,[13] Chapter, Lemma) on the uniform boundedness of a_n, as $n \to \infty$. Then

$$\int_{d_n \leq |\lambda - \mu|} K_n^2(\lambda, \mu) d\lambda d\mu \leq C d_n^{-2} = O(\log^{2/3} n),$$

which implies that

$$\text{Tr } K_n[\Delta](1 - K_n[\Delta]) = \int_a^{a+d_n} d\lambda \int_{a-d_n}^a d\mu K_n^2(\lambda, \mu) \qquad (33)$$

$$+ \int_{b-d_n}^b d\lambda \int_b^{b+d_n} d\mu K_n^2(\lambda, \mu) + O(\log^{2/3} n) = I_a + I_b + O(\log^{2/3} n).$$

To find I_a, we apply the results of,[4] according to which for λ, μ from the bulk of the spectrum the reproducing kernel has the form

$$K_n(\lambda, \mu) = h(\lambda, \mu) \frac{\sin n\pi(\phi(\lambda) - \phi(\mu))}{\pi(\lambda - \mu)} (1 + O(n^{-1})) \qquad (34)$$

$$+ \sum_{\pm} r_{\pm, \pm}(\lambda, \mu) e^{i\pi n(\pm\phi(\lambda)\pm\phi(\mu))},$$

where h and ϕ for (λ, μ) in the bulk of the spectrum are smooth, positive, bounded from both sides functions, the remainder functions $r_{+,+}$, $r_{-,-}$, $r_{+,-}$, $r_{-,+}$ have uniformly bounded derivatives in both variables, and $\sum_{\pm}$ means the summation with respect to all combinations of signs in the exponents. Moreover,

$$\phi'(\lambda) > c_0, \quad \text{if} \quad |\lambda - E_k| \geq \tilde{\varepsilon}, \quad k = 1, \dots, 2q,$$
$$h(\lambda, \lambda) = 1.$$

It is easy to see that the remainder terms in the r.h.s. of (34) after integration in the limits, written in the r.h.s. of (33), give us at most $O(d_n^2)$. Hence, we need only to find the contribution of the first term of (34). Performing the change of variables $\lambda = a + x/(n\phi'(a))$, $\mu = a - y/(n\phi'(a))$, we get

$$I_a = \int_0^{nd_n} dx \int_0^{nd_n} dy (1 + O(1)) \frac{\sin^2(\pi(x+y)(1+O(1)))}{\pi^2(x+y)^2} dxdy + O(d_n^2)$$

$$= \int_0^{nd_n} dx \int_0^{nd_n} dy \frac{\sin^2(\pi(x+y))}{\pi^2(x+y)^2} dxdy + O(\log nd_n) + O(d_n^2)$$

$$= \left(\int_0^1 dx + \int_1^{nd_n} dx \right) \left(\int_0^1 dy + \int_1^{nd_n} dy \right) \frac{\sin^2(\pi(x+y))}{\pi^2(x+y)^2} dy +$$

$$+ O(\log nd_n) + O(d_n^2)$$

$$= \int_1^{nd_n} dx \int_1^{nd_n} dy \frac{1 - \cos 2\pi(x+y)}{2\pi^2(x+y)^2} + O(1) + O(\log nd_n) + O(d_n^2)$$

$$= \int_1^{nd_n} dx \int_1^{\infty} dy \frac{1}{2\pi^2(x+y)^2} + O(1) + O(\log nd_n)$$

$$= \frac{1}{2\pi^2} \log n(1 + O(1)).$$

Similarly

$$I_b = \frac{1}{2\pi^2} \log n(1 + O(1)).$$

Then, in view of (33), we obtain (29).

□

Proof of Theorem 2
Let us consider the eigenvalue problem for $\hat{K}_{n,1}[\Delta]$:

$$\begin{cases} S_{n,1}f_\Delta + D_{n,1}g_\Delta = Ef_\Delta, \\ I_{n,1}f_\Delta - \epsilon f_\Delta + S_{n,1}^T g_\Delta = Ef_\Delta. \end{cases} \tag{35}$$

Here and below

$$f_\Delta = 1_\Delta f, \quad g_\Delta = 1_\Delta g.$$

Observe that since all the functions in the first line of (35) are analytic, the equation is valid also outside of Δ. Apply the operator ϵ to both sides of the equation. We get

$$\begin{cases} I_{n,1}f_\Delta + S_{n,1}^T g_\Delta = E\epsilon f_\Delta + E\epsilon f_{\bar{\Delta}}, \\ I_{n,1}f_\Delta - \epsilon f_\Delta + S_{n,1}^T g_\Delta = Eg_\Delta, \end{cases} \tag{36}$$

$$\Rightarrow Eg_\Delta = (E-1)(\epsilon f_\Delta) + 1_\Delta E\epsilon f_{\bar{\Delta}},$$

where we use that integration by parts gives us that $\epsilon D_{n,1} = S_{n,1}^T$, and denote

$$f_{\bar{\Delta}} = f - f_\Delta$$

with $\bar{\Delta}$ being a complement of Δ. Observe that

$$\epsilon f_{\bar{\Delta}}(\lambda) = \frac{1}{2} \int_{-\infty}^{a} f(t)dt - \frac{1}{2} \int_{b}^{\infty} f(t)dt =: (f, \Psi_{\Delta}) = \text{const}, \quad \lambda \in \Delta. \quad (37)$$

Multiply the first line of (36) by E and use the above equation for Eg_{Δ}. Integration by parts gives us

$$\mathcal{D}_{n,1}(\epsilon f_{\Delta})(\lambda) = \mathcal{S}_{n,1} f_{\Delta}(\lambda),$$
$$\mathcal{D}_{n,1}(1_{\Delta} \epsilon f_{\bar{\Delta}})(\lambda) = (\mathcal{S}_{n,1}(\lambda, a) - \mathcal{S}_{n,1}(\lambda, b))(f, \Psi_{\Delta}) =: Pf, \quad (38)$$

where by the above definition, P is a rank one operator in $L_2[\Delta]$. We obtain

$$(2E - 1)\mathcal{S}_{n,1}[\Delta]f_{\Delta} - E^2 f_{\Delta} + EPf = 0.$$

Hence the solutions $\{E_k\}$ of (35) are solutions of the equation

$$\mathcal{P}(E) := \det\left\{E^2 - (2E - 1)\mathcal{S}_{n,1}[\Delta] + EP\right\} = 0.$$

It is evident that $\mathcal{P}(E)$ is a polynomial of $2n$th degree, and E_k are the roots of $\mathcal{P}(E)$. We are interested in

$$\prod_{k=1}^{2n}(1 + \delta_n E_k) = \delta_n^{2n} \prod_{k=1}^{2n}(\delta_n^{-1} + E_k) = \delta_n^{2n}\mathcal{P}(-\delta_n^{-1}),$$

where $\delta_n := e^{x_n} - 1$. Thus we obtain

$$\log \hat{\Phi}_{n,1}(x) = -x_n E\{\mathcal{N}_n[1_{\Delta}]\} + \frac{1}{2}\log\det\left\{1 + (2\delta_n + \delta_n^2)\mathcal{S}_{n,1}[\Delta] + \delta_n P\right\}. \quad (39)$$

Now we use (22). Substituting the representation in (39) we get

$$\log \hat{\Phi}_{n,1}(x) = -x_n E\{\mathcal{N}_n[1_{\Delta}]\} + \frac{1}{2}\log\det\left\{1 + (2\delta_n + \delta_n^2)K_n(\Delta)\right\} \quad (40)$$
$$+ \frac{1}{2}\log\det\left\{(1 + R(\delta_n P_1 + \tilde{\delta}_n n \sum_{k,j=-(2m-1)}^{2m-1} F_{kj}^{(1)}Q_{kj}\right\},$$

where $\{Q_{kj}\}$ are rank one operators with the kernels $Q_{kj}(\lambda, \mu) = \psi_{n+k}(\lambda)\epsilon\psi_{n+j}(\mu)$,

$$R = (1 - \tilde{\delta}_n K_n[\Delta])^{-1}, \quad \tilde{\delta}_n = (e^{2x_n} - 1) = 2\delta_n + \delta_n^2.$$

According to the standard linear algebra argument

$$\det(1 + \tilde{\delta}_n \sum a_i \otimes b_i) = \det\left\{\delta_{ij} + \tilde{\delta}_n(a_i, b_j)\right\}.$$

Taking into account the formula and the structure of the remainder in (40), we conclude that in order to prove that the last term in (40) is small, it suffices to prove that

$$|n(R\psi_{n-j}, \epsilon\psi_{n+k})| \leq C\delta_n, \tag{41}$$
$$|(R1_\Delta S_n(x, a), \Psi_\Delta)| \leq C\delta_n, \quad |(R1_\Delta S_n(x, b), \Psi_\Delta)| \leq C\delta_n.$$

The last two inequalities are trivial, since

$$\operatorname{supp} \Psi_\Delta = \bar{\Delta}, \quad \operatorname{supp} R1_\Delta S_n(x, a) = \operatorname{supp} R1_\Delta S_n(x, b) = \Delta.$$

For the proof of the first inequality of (41) we need the following lemma.

Lemma 2. *Set*

$$v_n(\lambda) := 1_{\bar{\Delta}}(\lambda) \int_\Delta d\mu K_n(\lambda, \mu). \tag{42}$$

Then for any $\lambda \in \bar{\Delta}$

$$|v_n(\lambda)| \leq \frac{C}{1 + n|\lambda - b|} + \frac{C}{1 + n|\lambda - a|}. \tag{43}$$

The proof of the lemma is given after the proof of Theorem 2. Now we continue the proof of (41). The first bound of (41) is a corollary of three estimates

$$|n(1_\Delta\psi_{n+k}, \epsilon\psi_{n+j})| \leq C\delta_n, \quad |n(\psi_{n+k}, K_n[\Delta]\epsilon\psi_{n+j})| \leq C\delta_n, \tag{44}$$
$$\|n(K_n[\Delta] - K_n^2[\Delta])\epsilon\psi_{n+j})\| \leq C.$$

Indeed, the third bound of (44) yields for $m \geq 2$

$$\|n(K_n^m[\Delta] - K_n[\Delta])\epsilon\psi_{n+j})\| \leq \sum_{l=0}^{m-1} \|K_n^l[\Delta] \, n(K_n^2[\Delta] - K_n[\Delta])\epsilon\psi_{n+j}\|$$

$$\leq m\|n(K_n^2[\Delta] - K_n[\Delta])\epsilon\psi_{n+j})\| \leq mC.$$

Here we used also that $\|K_n[\Delta]\| \leq 1$. Thus,

$$\left\| \sum_{m=2}^\infty \tilde{\delta}_n^m n(K_n^m[\Delta] - K_n[\Delta])\epsilon\psi_{n+j}) \right\| \leq \sum_{m=2}^\infty mC\tilde{\delta}_n^m \leq C\tilde{\delta}_n^2,$$

$$\Rightarrow \left\| \sum_{m=2}^\infty n\tilde{\delta}_n^m K_n^m[\Delta]\epsilon\psi_{n+j} - \frac{n(2\tilde{\delta}_n^2 - \tilde{\delta}_n^3)}{(1 - \tilde{\delta}_n)^2} K_n[\Delta]\epsilon\psi_{n+j} \right\| \leq C\tilde{\delta}_n^2.$$

Combining this inequality with the first two bounds of (44) we obtain the first bound of (41).

To prove (44), we use the result of [15, Lemma 2], according to which

$$\epsilon \psi_{n+j} = n^{-1/2} c_{n+j} + O(n^{-1}),$$

where c_{n+j} is some constant, bounded uniformly in n. Using this fact, we conclude that to prove (44) it suffices to prove that

$$|n^{1/2}(1_\Delta \psi_{n+k}, 1_\Delta)| \le C\delta_n, \quad |n^{1/2}(K_n[\Delta]\psi_{n+k}, 1_\Delta)| \le C\delta_n, \qquad (45)$$
$$\|n^{1/2}(K_n[\Delta] - K_n^2[\Delta])1_\Delta)\| \le C.$$

Lemma 2 yields

$$\|(K_n[\Delta] - K_n^2[\Delta])1_\Delta\| = \left\| \int_{\bar\Delta} d\nu \int_\Delta d\mu K_n(\lambda,\nu) K_n(\mu,\nu) \right\|$$
$$= \|K_n v_n\| \le \|v_n\| \le C_1 n^{-1/2},$$

hence we obtain the last inequality of (45). The first inequality of (45) is a simple corollary of the result of [4, Theorem 1.1], according to which

$$\psi_{n+k}(\lambda) = R_k(\lambda) \cos(n\pi\phi(\lambda) + m_k(\lambda)) \left(1 + O(n^{-1})\right)$$

where R_k and m_k are smooth functions. Using this result, we can integrate by parts and obtain the first inequality of (45) (even with $Cn^{-1/2}$ in the r.h.s. instead of $C\delta_n$). In addition, since

$$K_n \psi_{n+k} = 1_{k<0} \psi_{n+k},$$

we have

$$K_n[\Delta]\psi_{n+k} = 1_{k<0} 1_\Delta \psi_{n+k} - K_n 1_{\bar\Delta} \psi_{n+k}.$$

The bound of the first term is given by the first inequality of (45). For the second term write

$$|(K_n 1_{\bar\Delta} \psi_{n+k}, 1_\Delta)| = |(1_{\bar\Delta} \psi_{n+k}, K_n 1_\Delta)| = |(1_{\bar\Delta} \psi_{n+k}, v_n)| \le Cn^{-1/2}.$$

Hence, we complete the proof of the second inequality of (45).

It was explained above that (45) imply (44), which combined with (40) yields

$$\log \hat\Phi_{n,1}(x) = -x_n E\{\mathcal{N}_n[1_\Delta]\} + \frac{1}{2}\log\det\left\{1 + (e^{2x_n} - 1)K_n[\Delta]\right\} + O(\delta_n).$$

Then similarly to the case $\beta = 2$ we have

$$\frac{1}{2}\mathrm{Tr}\,\log(1 + (e^{2x_n} - 1)K_n[\Delta]) = x_n E\{\mathcal{N}_n[1_\Delta]\} + x_n^2 \mathrm{Tr}\,K_n[\Delta](1 - K_n[\Delta])$$

$$+ \frac{(2x_n)^3}{12}\mathrm{Tr}\,K_n[\Delta](1 - K_n[\Delta])\tilde R(K_n[\Delta]). \qquad (46)$$

By Lemma 1,

$$\operatorname{Tr} K_n[\Delta](1 - K_n[\Delta]) = \pi^{-2} \log n \, (1 + o(1)),$$

hence the limit of the second term of (46) is x^2 and the last term is $O(\log^{-1/2} n)$.

$\square$

Proof of Lemma 2 The proof is based on the representation (34). Integrating by parts, it is easy to see that the contribution of the remainder terms (written in $\sum_{\pm}$) is at most $O(n^{-1})$. Hence we need to consider only the contribution of the first term in the r.h.s. of (34). Take $\lambda < a$ and consider the change of variables $x = \phi(a) - \phi(\lambda)$, $y = \phi(\mu) - \phi(a)$. Let φ be the inverse function of $\phi(x) - \phi(a)$ and $a - \lambda = \Delta_\lambda \geq 0$. Then the main part of our integral takes the form

$$F(\lambda) := \int_0^d dy \varphi'(y) \tilde{h}_\lambda(y) \frac{\sin n\pi(x + y)}{\varphi(y) + \Delta_\lambda} \tag{47}$$

$$= \varphi'(0) \tilde{h}_\lambda(0) \int_0^d dy \frac{\sin n\pi(x + y)}{\varphi(y) + \Delta_\lambda} + O(n^{-1})$$

$$= C \int_0^{nd} dy' \frac{\sin \pi(nx + y')}{n\varphi(y'/n) + n\Delta_\lambda} + O(n^{-1}),$$

where

$$d = \phi(b) - \phi(a), \quad \tilde{h}_\lambda(y) = h(\lambda, \varphi(y)),$$

and we have used the fact that the function

$$\frac{\varphi'(y)\tilde{h}_\lambda(y) - \varphi'(0)\tilde{h}_\lambda(0)}{y + \Delta_\lambda}$$

has a bounded derivative, hence integration by parts with $\sin n\pi(x + y)$ gives us $O(n^{-1})$

$$F(\lambda) := C \int_0^{1-\{nx\}} dy \frac{\sin \pi(\{nx\} + y)}{n\varphi(y'/n) + n\Delta_\lambda}$$

$$- C \sum_{k=1}^{nd} \left(\int_{-\{nx\}}^{1-\{nx\}} dy \frac{\sin \pi(\{x\} + y)}{n\varphi((y' + k)/n) + n\Delta_\lambda} \right.$$

$$\left. - \int_{-\{nx\}}^{1-\{nx\}} \frac{\sin \pi(\{nx\} + y')}{n\varphi((y' + k + 1)/n) + n\Delta_\lambda} dy' \right).$$

Observe that the series above is of alternating sign, and modules of the terms decay, as k grows (recall that $\varphi(y)$ is an increasing function of y). Thus,

$$F(\lambda) \geq C \int_0^{1-\{nx\}} dy' \frac{\sin \pi(\{nx\} + y')}{n\varphi(y'/n) + n\Delta_\lambda}$$
$$- C \int_{-\{nx\}}^{1-\{nx\}} dy \frac{\sin \pi(\{nx\} + y)}{n\varphi((y'+1)/n) + n\Delta_\lambda},$$

$$F(x) \leq C \int_0^{1-\{nx\}} dy \frac{\sin \pi(\{nx\} + y)}{(x+y)}$$
$$- C \int_{-\{nx\}}^{1-\{nx\}} dy' \frac{\sin \pi(\{nx\} + y')}{n\varphi((y'+1)/n) + n\Delta_\lambda}$$
$$+ C \int_{-\{nx\}}^{1-\{nx\}} dy' \frac{\sin \pi(\{nx\} + y')}{n\varphi((y'+2)/n) + n\Delta_\lambda}.$$

These bounds combined with (47) prove (43) for $\lambda < a$. For $\lambda > b$ the proof is the same.

$$\square$$

Proof of Theorem 3 The proof is very similar to that of Theorem 2, hence we present it very briefly. We consider the eigenvalue problem for $\hat{K}_{n,4}[\Delta]$ ($\beta = 1$).

$$\begin{cases} \mathcal{S}_{n,4}f_\Delta + \mathcal{D}_{n,4}g_\Delta = Ef_\Delta, \\ \mathcal{I}_{n,4}f_\Delta + \mathcal{S}_{n,4}^T g_\Delta = Ef_\Delta. \end{cases} \tag{48}$$

Apply the operator ϵ to both sides of the first equation and then subtract the second line from the first. We get

$$Eg_\Delta = E(\epsilon f_\Delta) + 1_\Delta E \epsilon f_{\bar{\Delta}}.$$

Substituting the relation in the first line of (48), we obtain

$$2\mathcal{S}_{n,4}[\Delta]f_\Delta - Ef_\Delta + Pf = 0,$$

where (cf. (38))

$$Pf := (\mathcal{S}_{n,4}(\lambda, a) - \mathcal{S}_{n,4}(\lambda, b))(f, \Psi_\Delta)$$

is a rank one operator. Taking into account (19) and (15), we have now (cf. (39))

$$\log \hat{\Phi}_{n,4}(x) = -x_n E\{\mathcal{N}_n[1_\Delta]\} + \frac{1}{2} \log \det \left\{ 1 + \delta_n \mathcal{S}_{n,4}\Delta] + \frac{\delta_n}{2}P \right\}.$$

Applying (22) and repeating the argument used in the proof of Theorem 2, we obtain the assertion of Theorem 3.

References

1. A. Boutet de Monvel, L. Pastur, M. Shcherbina, *J. Stat. Phys.* **79**, 585 (1995).
2. Z. Bao, G. Pan, W. Zhou, *J. Stat. Phys.* **150**(1), 88 (2013).
3. O. Costin and J.L. Lebowitz, *Phys. Rev. Lett.* **75**, 69 (1995).
4. P. Deift, T. Kriecherbauer, K. McLaughlin, S. Venakides, X. Zhou *Commun. Pure Appl. Math.* **52**, 1335 (1999).
5. P. Deift, D. Gioev, *Int. Math. Res. Papers.* **004**, 116 (2007).
6. P.J. Forrester, J.L. Lebowitz, arXiv:1311.7126v3.
7. K. Johansson, *Duke Math. J.* **91**, 151 (1998).
8. M.L. Mehta, *Random Matrices.* New York: Academic Press (1991).
9. V. Kargin, *J. Stat. Phys.* **154**, 681 (2014).
10. T. Kriecherbauer, M. Shcherbina, arXiv:1003.6121.
11. L. Pastur, M. Shcherbina, *J. Stat. Phys.* **86**, 109 (1997).
12. L. Pastur, M. Shcherbina, *J. Stat. Phys.* **130**, 205 (2007).
13. L. Pastur, M. Shcherbina. In AMS, 634 pp., SURV/171 (2011).
14. M. Shcherbina, *Comm. Math. Phys.* **285**, 957 (2009).
15. M. Shcherbina, *Comm. Math. Phys.* **307**, 761 (2011).
16. M. Shcherbina, *J. Stat. Phys.* **151**(6), 1004 (2013).
17. M. Shcherbina, *J. Math. Phys.* **55**(4), 043504 (2014).
18. A. Soshnikov, *J. Stat. Phys.* **100**, 491 (2000).
19. C.A. Tracy, H. Widom, *J. Stat. Phys.* **92**, 809 (1998).
20. H. Widom, *J. Stat. Phys.* **94**, 347 (1999).

Printed in the United States
By Bookmasters